Headwaters
of the
River of Bullshit

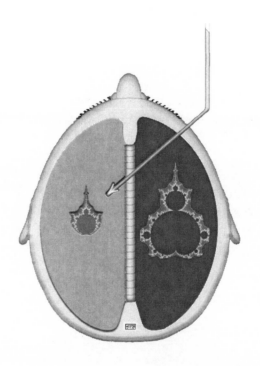

Alan F. Boyne, Ph.D.

To Dennis
With Thanks!

Alan F. Boyne

The Headwaters of the River of Bullshit

pISBN 978-1-4507-9063-5 (Print edition)
eISBN 978-1-62620-081-4 (Multimedia Edition)

W. Bruce Conway—Publisher
1st Printing April 2013
Revised: May 2013

For information address:
Alan Boyne, P.O. Box 548, Friday Harbor, WA 98250

Library of Congress information is available from the publisher upon request.

Includes bibliographical references and index.

File Under: Psychology, Memoir, Science

Website: http://www.River-Of-Bull.com

Dedication

For all those who have ever wondered
why things don't turn out as well as they could.

Preface

This book is the narrative of an actual life—mine—in which I use modern brain science to address a particular psychological problem that I first recognized while growing up in the docklands of Liverpool.

The alleged ability of my people to talk the hind leg off a donkey was apparently developed during the pre-electronic time, when there was a need for entertainment over the evening meal and funny accounts of the day's events were the natural raw material. Long in place amongst the Irish, the practice has a name: *blarney*, and I have enjoyed my fair share.

One day, however, I saw the borderline across which *blarney* morphs into *bullshit*. Bull typically delivers a brief ego boost but it leaves the future hostage to unresolved conflicts with reality. Prime example: an overheated planet.

In my mid-twenties, already convinced that the human adventure was bogged down in swampy bottomlands of rationalization, I stumbled across an opportunity to become a neuroscientist. As I contemplated the option, it suddenly seemed obvious that this new discipline was likely to discover the root of the blarney/bullshit dichotomy and civilization's associated discontents. In 1967, as I scribed my name on the dotted line, I imagined that, when the great day dawned, there would be a handshake of global proportions after which we would all set about draining the swamp.

Good news: Although it took longer than I had expected, the exploration work has been done, and the credits are established (none of which redound in my direction). Consequently, for more than a decade past, there has been a potential to kindle a firestorm of spiritual renewal, one that might rapidly enable broad cooperation across a multitude of parochial fault lines, and so turn back the tide of global degradation.

Not so good news: Science funding and laboratory practice encourage reductionism, and our reporting industry seems relatively incapable of framing the big picture in a manner that would either entertain at the evening meal or support a public-works project of the requisite magnitude.

Although I will attempt to fill this void in two books, "I yam what I yam," as Popeye would say, and that includes strong consciousness of both my working class origin and a relative lack of literary sophistication. As I consider this, the final version of the first book, however, I am happily reminded of the notebooks that I enjoyed creating in the days when *The Beatles* and *The Rolling Stones* played for our spring dances at Liverpool University.

Despite inevitable flaws and mistakes yet to be discovered, early feedback suggests that, at least for some people, *Headwaters* delivers the intended epiphanies.

Contents

Part Two: The Silent Witness

Kindred Books

(In reverse chronological order)

Cosmos Consciousness: The Survival Value of Empathy
Alan F. Boyne (Pending)

Tug Boat Man
Thomas W. Phillips (2013)

The Righteous Mind
Jonathan Haidt (2012)

The Republican Brain
Chris Mooney (2012)

The Master and his Emissary;
The Divided Brain and the Making of the Western World
Iain McGilchrist (2010)

The Political Mind:
Why you can't understand 21st-century American politics with
an 18th-century brain
George Lakoff (2008)

TheAlphabet versus the Goddess:
The Conflict between Word and Image
Leonard Shlain (1998)

The User Illusion: Cutting Consciousness Down to Size
Tor NØrretranders (1998)

The Origins of Consciousness in the
Breakdown of the Bicameral Mind
Julian Jaynes (1976)

The Behavior of the Lower Organisms
Herbert Spencer Jennings (1911)

Introduction

The Division of Mental Labor

The Left Hemisphere

Our experience of self-consciousness and a silent inner dialogue can be traced to the left hemisphere of most right-handers. When one asks out loud what another thinks, and a reply returns, both individuals are speaking and translating language with their left hemispheres. (This easy readout of what happens inside the head is so effective you might wonder whether brain scanners are worth all the money they cost.)

The Right Hemisphere

But when it comes to the right hemisphere, the situation is entirely different. Questioned all the livelong day, it will behave like the tar baby in the old plantation story—and say not a word. And yet brain scanners report that it is busily doing something.

Is it just mute? It cannot be described as simply mute, for you may overhear it singing beautifully in the shower. Furthermore, the right side of the brain can recite narrative poetry, and even swear! In response to these seemingly contradictory qualities, the egotistical and very status-sensitive left brain may jump to a Brer Rabbit conclusion: *"Der rite hems'feer is jus' plain ornery!"*

1

More academic forms of disparagement have included the notion that the right got left behind when language evolved; that it is now an animalistic relic of life on the other side of an evolutionary railroad track. One comic even suggested that it was a counterweight, that without it, our heads would flop onto our left shoulders.

I will argue that all such derision has crossed the line into—I hope you guessed it—bullshit. Moreover, I believe that now is a good time to step back and start the story of the brain over again, using a pair of recent clues that surely invite a collective dope slap: *Why didn't we pick up on this sooner?*

The Dichotomy

We have learned that evolution long ago established the survival value of focusing separately on details in the left cerebral hemisphere and their context in the right.

If natural selection gave roughly equal neuronal weight to each of these, perhaps we humans, reacting to a recently evolved ability to verbalize thought sequences, too readily plunge into obsession with details of linear cause and effect, of short-term profit and loss. Perhaps we have underweighted complex, multi-dimensional context considerations—such as the effect we have been having on our home planet.

When the silent side of the brain is fully credited with the valuable purpose of context comprehension, one can start the story of the brain over again, for it can then—and only then—be interpreted as a creative rationality engine with a built-in potential for mutual crosschecks. Once this and related conceptions are in place, it becomes evident that we have indeed been suffering the consequences of not having an adequate instruction manual.

Moreover, the unfolded perspective from neuroscience explains the relationship between creative blarney and delusional bullshit, the very Dog Star of a puzzle that guided my life's journey.

My multimedia notebook approach caters to each hemisphere, for the left reads writing, while the right better absorbs pictures. The aim is to encourage the reader to 'feel' both systems of perception, as well as the process of switching between them, until an epiphany takes place, at which point the right hemisphere usually roars with delight—a tar baby no more.

Part One

The Journey

Chapter 1
The Liverpool Docklands

The first of Liverpool's slaving ships, Liverpool Merchant, left wooden wharves at the Mersey shore and set sail for the coast of Africa on the 3rd of October 1699. When her cargo was sold in the Americas, the burghers of the small fishing community were rewarded, if that's the right word, with substantial profits.

Soon after that, the legendary "Liver birds" (perhaps cormorants) were chased out of the local tidal pool, for their nesting grounds were the best place to install masonry wharves, from which to harvest more profits. Subsequent wet docks were equipped with locks, so that tidal conditions imposed no restrictions on loading and unloading cargo.

By 1807, with a century of the slave trade behind her, Liverpool's population had soared; she was now a city with miles of the finest docklands in the world, handling forty percent of Europe's and eighty percent of Britain's slave trade, as well as an extensive transport in cotton and manufactured goods. An economic miracle!

Liverpool had thus become second only to London as a financial capital. But 1807 was the year in which official sanction for the slave trade was withdrawn. No matter; nudge—wink; the slaving ships raised foreign flags, and so an unofficial trade continued. Another quarter of a century passed before moral repugnance enforced an end to the supply. Shortly after that, the American Civil War forced an end to the demand. From that time forward, Liverpool's docklands experienced a long, slow decline, for other ports could handle unshackled cargo with competitive efficiency.

In 1939, Adolf Hitler's sense of entitlement to national acreage provoked Germany into invading the rest of Europe. Britain declared war and was soon besieged by sea and air. She was fortunate that Liverpool's marine stonehenge could still anchor a lifeline of convoy ships bringing food and munitions from America. I was born in the middle of The Second World War, near the Gladstone Docks, in the suburb called Bootle, which had been of considerable interest to the Luftwaffe. Consequently, I have a collection of stories about the funny side of being bombed out of house and home, some of which I will tell you in this chapter; you'll enjoy being horrified by them.

Before I was three years old, my island home had served as a launching pad from which the Allies had dismantled Nazi rationalizations. Since then, I have spent my allotted years trying to get some sort of intellectual grip on the human proclivity for self-serving bullshit. In keeping with a hardscrabble family background, I have an instinct for the jugular that seems to have helped. So now let me tell you how my parent's lives shaped mine.

Isabelle

Belle, my mother, was the youngest of six daughters. Her father had been a french polisher in the Cammell Laird shipyards on the Birkenhead side of the River Mersey. In a submarine renovation project during WWI, he was assigned to paint the interiors.

6

Although he was prone to bronchitis, nobody foresaw the effect of paint fumes in confined quarters. He died within two years of being diagnosed with irreversible lung damage.

My widowed grandmother then took her daughters, three of whom were of working age, across to the Liverpool side of the river, where they supported themselves by running a boarding house for sailors. Reporting on the years before she was born, my mother's older sisters described a delightful family. But for Belle, the only home she would remember was the boarding house. Nevertheless, she was never lonely there, and remembered coming home from school to play games and listen to stories from its forty-odd tenants.

While mother was growing up just outside the Gladstone Docks, my father-to-be, Fred, was being raised next door. His dad, Andy, worked for British Rail. In this relatively elevated position, Andy Boyne would become the only man on the street who received a pension at the end of his working life.

When Fred was eleven, his mother Mary immigrated to Australia, taking Fred and three siblings with her. Father Andy stayed behind because he had two more years to do to get that pension vested. As I will explain in the next chapter, Fred later found his way back to the boarding house, and eventually married Belle. This blissed-out image was made during courting days, before the trauma of September in 1939 when Nazi forces invaded Poland, and England responded by declaring war.

Already a merchant mariner, Fred was obliged to sail on the convoy ships. Although he survived, the lives of many of his friends ended when torpedoes tapped against ship's hulls in mid Atlantic. Each mind, the fruit of nearly fourteen billion years of evolution, had thus been put to no better use than victimhood in mortal combat.

When Fred had first gone to sea, he was a hungry refugee from the Great Depression. But with the help of ship's food and Charles Atlas bodybuilding techniques, he developed a muscular torso and became a prize-winning boxer and occasional wrestler. His ring name was 'Gentleman Jim' (after a pugilist of an earlier time).

I loved him and thought he was the strongest man in the world. Had he opened a can of spinach with a one-handed squeeze, I would have considered it quite natural. I retouched the above figure from a Kodak Brownie camera snapshot of those days. Note the rope belt. This image perfectly captures my childhood conception of my father.

The Canary Bird

The saga of the canary bird happened on a spring night before I was born. An Atlantic convoy had returned safely and Fred, recently married to Belle, was living with her extended family in the boarding house, just outside Gladstone Docks.

Near midnight, air-raid sirens warned that the Luftwaffe was approaching. Both the docks themselves and the homes of the dockworkers were targets. On the first night, a three-story pub took a direct hit from a 500-pounder, after which no walls remained. There was no roof either, of course, and even the pile of rubble on the ground seemed irrationally small. In view of such effects, the air-raid sirens were the signal to head for the bomb shelters. These were corrugated iron arches, set into the soft earth at the bottom of the back garden. It was cold and damp down there, but a safer space than a house of bricks.

When the "All Clear" sounded, Fred's buddy, "Chinny" Biggins, rousted Fred from the bomb shelter to survey the pub. They remembered that the proprietor had planned to take cover below ground, among the beer barrels in the cellar. All available hands began removing rubble from the cellar entrance.

Sure enough, the publican emerged like a modern Lazarus, along with his whole family. In the emotional algebra of the time, the Luftwaffe lost because someone survived a direct hit. With no home, business, or possessions, however, the pub family took a bus to "the country" and never returned.

Recordings of Winston Churchill's gravel-toned wartime radio addresses are still available, and one can readily imagine Winnie's cultivation of defiance in every corner of the British Empire, and so saving the nation. But at the moment, I have the saving of a particular little bird in mind.

Despite the shelters, the death toll was substantial, and daylight revealed more corpses than could be buried before night fell again. To make matters worse, the Luftwaffe would bomb for seven nights in a row. In some desperation, the public swimming baths at Marsh Lane were emptied, and the unfortunate dead were piled within its tiled spaces until the carnage ended. The final death toll was 1740, and 550 of these unknown, presumably unidentifiable "warriors" were buried in a mass grave at Anfield cemetery.

The worst of the bombing occurred on the third night. But when everyone reached the shelter, Granny realized that she had left the canary behind. Fred went back and did the rescue, just as the explosions began anew. That action was polished up for a permanent place in the collective memory. Gales of laughter still sound in my mind from postwar weddings and funerals, when those wartime experiences (that I was almost sorry to have missed) were recounted.

The Night the Mouse Died

More than fifty years later, I was searching to understand my place in both the local tapestry and the broader scheme of human events. On a visit home, I was tape-recording Mother's account of that May Blitz. She mentioned that the familiar whine of a descending bomb had an unusual character when it was aiming directly at your personal ears. "How the hell could anyone know that?" said I.

"I'm trying to tell you!" said Belle, "Shut up and listen!" She hadn't had a chance to say that to me for a long time. For six nights the extended family had made the midnight dash to bunk with the fairies at the bottom of the garden. On the seventh night, as Belle and Fred heard the sirens crank up their wailing yet again, they decided to let everyone else go to the shelter without them. As they lay abed, a descending bomb screamed through the sky... with a sound like no bomb they had ever heard.

They could only hold each other and wait to die. The house shuddered as the missile buried itself full deep into the earth of the back yard, three feet from the roofline. Still alive to the silence that followed, they realized that it had failed to detonate, and so the house had failed to disintegrate. Unexploded munitions were quite common, which provided the answer to my impatient question. Some such bombs went off anyway—after a delay—so the civilian guard bomb squad came and insisted upon rapid evacuation.

Once gathered outside the dock-wall, in the darkness of mandatory blackout, and long past the midnight hour, there was a what-the-hell-shall-we-do-now moment. Engines roared overhead; concussive explosions filled the night air; and the ground defenses drove their guns into a frenzy of ack-ack chatter. With shrapnel whining by, Belle shoved her hands into her coat pocket and felt her fingers wrap around the keys to Barry's Tea House. This was run by her sister, Winnie, and catered to the dockworkers; it was only a half-mile away. The family set off by the light of burning buildings. In my mind's eye, I see one of them carrying a birdcage.

Opening the front door, Belle felt her way along a hallway to the kitchen, where she had to strike a match to light the gas lamp. As the mantle ignited, a mouse, harvesting crumbs on the table, came into full view, which prompted an involuntary bloodcurdling scream (from Mother, not the mouse). After seven nights of being bombed in the dark, a final experience of being screamed at in the light was more than this house mouse could endure. It leaped high in the air and landed back on the table, stone dead.

We will be soon discussing the ability of the brain to make loose associations. This story is perhaps my favorite example, for the complete set of associations of that awful night, in the minds of all who lived it, could be evoked by a simple question: "Do you remember the night that the mouse died?"

I learned that I had been born nine months after the spring night that the mouse died.

My Time Begins

When I check my personal narrative, of the earliest things I can remember, the only war trauma I recall concerns our diet. We ate scrounged-together delicacies called scouse and hotpot, and occasionally had tripe (the lining of a cow's stomach). I hated the tripe and merely disliked the other two. Now that I think of it, the little flat bottles of cod liver oil delivered by the National Health Service were also traumatic. Of course the oil was a valuable nutritional supplement, but it was still extract of dead fish.

I preferred a second flat bottle, which extended old British Navy wisdom: limes keep sailors free of scurvy on long voyages, and orange juice serves for landlubbers during World Wars.

On an occasional Sunday when Fred was home from the sea, and I must have been about four or five, he would sit me on the crossbar of his bike and ride out to the countryside, to the Pheasant Inn.

Those adventures, still gleaming in memory, seem to reflect the way the world should be. Children were not allowed in pubs, but Fred was able to bring out two pints of Guinness to a ditch on the other side of the road. I would hold one while he drank the first and told me stories. As we sat in the grass, we could see a concrete pillbox, a leftover from the recent war, in the distance. Positioned at the edge of a farmer's field, about a mile from the Irish Sea, its gun slit covered the road from the beach.

On the other side of the road a dense pine forest grew out of a carpet of shiny brown needles. While clutching 'my' pint, I learned that the mysteriously gloomy forest was the wood I had heard of in a fairy tale. It was the actual Sir Percy's Wood! I was wide-eyed with childish astonishment, gobsmacked, as the Irish say.

Having captured my attention, Dad went on to explain that King Arthur and Sir Percy remained in the wood, awaiting England's need. If the Nazis had invaded, Merlin would have released them to do battle. How perfectly appropriate, I must have thought, peering into the shadows among the trees, hoping to catch the flash of a lance, a heraldic ribbon, or maybe a big black horse.

As my glance returned to the pillbox, Fred brought the tale up to date. Apparently someone in the British army had realized, he said, that knightly lances and chain mail would no longer be good enough, so they had provided a concrete redoubt and a machine gun. Although I preferred to imagine the great king astride a rampant stallion—and it was hard to see how the rest of them were going to fit in there—I was too young and impressionable to register misgivings.

In fact, I recounted the tale to Belle the next morning at breakfast, as though it was valuable received wisdom. She rolled her eyes and said, "Your father couldn't just kiss the Blarney Stone like anyone else; he had to swallow the damn thing!" This unique stone is in Blarney Castle, Ireland. When kissed, the stone was supposed to confer the ability to talk a girl into marriage. I can't be sure, but this rebuke of my father's story may have been my first chance to recognize creative blarney.

Fred had qualified as a rigger and was comfortable with rope. Whenever anything broke at home, he would announce with evident satisfaction that he could fix it with a bit of string—the universal precursor of duct tape. One day, Belle declared that Fred had an Indian name that I ought to be aware of. "What's that?" I asked. "Bitter String," she replied. It took me a while.

Despite his projection of physical competence, Fred's self-image had big holes in it. This only became apparent when I was older and noticed that everything practical that he attempted to do, i.e., projects requiring more than string, seemed to fail. I eventually realized that he had left school too early to learn the technological underpinnings of our rapidly advancing civilization.

To make matters worse, he developed a phobia about not doing things "right" the first time. When one is afraid to make the first mistake, one can't benefit from trial and error—nature's primary discovery tool. Our emotional bond, however, was strong and to this day I can sometimes feel a continuum between our lives— perhaps a consequence of resonating with imagined emotions as he described his life experiences. When he was on the beach, as sailors say, he would try to get day work at the docks. That meant being out of bed and into the lineup before dawn, hoping to be selected. Previous attention to your relationship with the foreman, who did the selecting, was essential. (As in, "Can I buy you a drink?")

The amount of money one earned was a further insult. Fred and his close friend, Billy, developed a payday routine of throwing their pay packets (cash in an envelope) on the ground. Then they tossed a coin to see who would take home both packets and who would go home with nothing. I'm sure the wives hated the practice, but I have no trouble understanding the need to defy that made Fred and Billy feign indifference to degrading circumstance.

If we understood our mental mechanisms better, we might be able to make economic degradation a thing of the past. But righteous egos, wielding copious bullshit, stand in the way

Manna from Heaven

Fred would make his pilgrimage to the docks before Belle and I got up in the morning. A golden memory formed around his habit of always leaving a hidden note somewhere in the kitchen. We could expect it to be funny, so we would eagerly empty our chamber pots in the outdoor toilet, and begin the ritual search.

One morning, when we had been keeping rabbits, Fred wrote that Albert rabbit had choked on an apple and died in the night.

Wondering what to do with the corpse, he suddenly remembered the Seaforth greyhound track, and the plight of the greyhounds chasing an electric rabbit on wire guides. Feeling sudden empathy for the existential angst of such a life, Fred decided that he would tuck Albert's body under his coat. He wrote that he was going to chuck him over the wall of the stadium as he cycled past.

The thought of greyhounds having their fantasies fulfilled, courtesy of my dad, tickled me endlessly; I just thought it was the best idea I ever heard of!

The hoary adult of today, however, has to suspect that we were keeping rabbits to compensate for the wartime meat shortage, and the whole tale was a joyful demonstration of the mind's ability to loose associate. At the evening meal, Baby Bear was probably eating Albert meat while giggling at the retelling of the tale, with Mother Bear and Father Bear laughing at Baby Bear's gullibility.

Mother

My mother was beyond my capacity to understand. One early Christmas morning, I had toddled into their bedroom to discover a black Scotty dog sandwiched between them.

He was evidently disappointed to learn of my existence, and could bark louder than me. He was sold forthwith to the local chip-shop as a certifiably aggressive guard dog, and we got a small mongrel puppy. Tinker and I bonded by chasing each other around the dustbin in the tiny backyard. Belle used the experience to explain that if I were troublesome, I could be disposed of in the same way, i.e., sold to the chip shop. This teasing might be put aside as being merely unfortunate but it was part of a pattern.

Once when I was three or four, she ordered me to leave home, and I was ushered outside the backyard door. When it closed behind me, I couldn't imagine where to go, and experienced a desperate feeling of total helplessness. I sat on the threshold stone and cried bitterly, etching the event permanently in my brain.

A stranger passed and asked why I was so upset. I explained that my mother had told me to leave and never come back, "... and I don't know where to go!" His instructions to return, and to tell my mother that he had sent me, provided the permission I needed to do so. I include this anecdote because doubts about motherly love may have contributed to the profound iconoclasm that I will later document.

Much later in life, Belle was diagnosed with Borderline Personality Disorder (BPD). Ironically, one characteristic of this syndrome is a pronounced fear of being left alone. In those days, an effort to seek psychiatric help was considered an admission of lunacy, and decades would pass before Belle met with an appropriate professional.

The first consultation, however, gave her nightmares and "That was it!" as she would say; she didn't return—but the nightmares did. In old age, they included recitations of all the wrongs she had experienced, and how she hated the people responsible. We were all on the list. The BPD condition is relatively unresponsive to therapy, so it's unlikely that any help would have come, had she been willing, or perhaps able, to seek it.

My brother was born when I was three and a half, and my sister when I was eight. The three of us were pawns in the developing battleground of our parent's marriage. Our survival can be attributed to nature's mechanisms, including the beneficial influence of an extended Irish family and to educational opportunities provided by the larger society. That brings me to schooling in Great Crosby, and the Penguin people.

The Village of Great Crosby

Every English schoolboy learns that, in 1066, a bloke called William, a Norman from France, invaded Britain, killed King Harold, and laid waste to the island nation. As was the tradition in those days, he declared himself the new king.

Twenty years later, he was broke and in need of more money—for defense of England against more people like him.

Along with belligerence, William had a more housebroken sense of bureaucracy, and His Majesty's Domesday Book was the first record of properties upon which taxes might be based. But when the king's men scoured the kingdom, twenty years after crushing all and sundry, they found that many places remained "wasteland" and could bring in no revenue. Ironies abound.

At the time of the great inscribing, a place called Crosby was already a settled piece of Crown Land. Nearby, however, the scribes found a small, unrecorded hamlet. Dubbing the place Little Crosby, they added it to the tax rolls. By my time, Great Crosby was a bustling township on the edge of the greenbelt that surrounds Liverpool. We moved there in about 1948.

On the outskirts of town, before one reaches Little Crosby, there is a small, forested area—a municipal wood. It was always referred to in the plural as Sniggery Woods and was surrounded by the ditches that irrigated the adjacent farm fields. This was a childhood paradise: fond memories remain of traveling its deeply rutted trails with Tinker, our scrappy little mongrel, and of straddling the ditches in pursuit of stickleback fish, tadpoles, and water beetles.

I don't know who the original Sniggery was, or how he managed to prosper with a name like that, but his legacy connected me to England's medieval past, provided blackberries for Mother's fabulous pies, and made the town an idyllic place in which to grow.

The Daily Odyssey

On weekdays, those of us children who lived on Victoria Road could put on school uniforms, strap on our leather satchels, and make an 8:30 a.m. trek toward the Mersey. The first landmark was a duck pond at old Wright's Pitt Farm. Around the far edge, rough stone farm buildings formed a rambling semicircle. Where the near bank abutted the pavement, wrought-iron railings kept us from falling in.

You could push your face between the iron bars—up to your ears—and peer at the black water and the tiddlers, the small fish that the ducks would dive for. We told ourselves that the pond had no bottom, for no trip to school was complete without telling ourselves stories, and we favored the most wonderful ones. (Since we were somewhat shy of four feet tall, a bottomless duck pond approximated reality.) The next element of excitement was the house with the boarded-up windows. It seemed best to run past, with heart pounding, for we were convinced that a witch lived there. After that came a wormhole of an underground tunnel that allowed pedestrians to continue an uninterrupted journey beneath the railroad tracks. This was Billy Goat Gruff territory.

I hated to be still in the tunnel when a train passed over us and the arched walls growled. More running was advisable. When we popped out on the other side, we were in Blundellsands, and the air felt different. A popular watering hole still stands right next to the train station, so it might have been beer barrel aroma that made the air different, but we didn't yet know about that. Not far away now the road ended at the shoreline of the Mersey. Masonry rubble from the many war-torn buildings had been dumped above the high tide line and then topped with black glassy material, some sort of industrial waste that sliced open the inquisitive index finger of a small boy. Just before the beach stood our daily destination: the home of mysterious black-robed nuns of an Ursuline Convent.

The Penguin People

Between ages six and seven I attended the convent, always eager to help Sister Barbara dispense the orange juice, for I had decided that she was nice, and that was exceptional. The buildings with their round, spired towers were set among sand dunes, sawgrass and magnificent glades filled with bushes and scented by bluebells. I'm disappointed with the piddly little things that get called bluebells today—you have to bend down to smell them.

Many years later, I learned that my educational circumstances had a history and a pseudo-rational explanation.

In an extension of the Conqueror's general attitude, Great Britain had eventually appointed itself to be in charge of a substantial portion of the "undeveloped world." The ostensible explanation was that the empire was bringing Christianity to the heathen masses, with an implication that the conversion of other people's natural resources into great wealth for England was simply an incidental result of 'being over there anyway.'

The heathen masses, referred to with a derogatory epithet that I should probably leave to the past, proved to be ungrateful wretches and, by the 1950s, the various constituencies of empire had followed America's lead and declared themselves independent.

Somewhere in the throes of this national decline, an epiphany struck: the leaders of not-so-great-anymore Britain realized that the upper classes weren't producing enough intelligent progeny to run one country properly, let alone much of the globe. It became necessary to seek out the brightest of the lower classes and train them to shoulder part of the burden. Unbeknown to myself, a bottle had spun in my direction.

At the age of seven or eight, we were separated from the young girls at the convent, and the old girls who ran it, and we were moved closer to the railway station, to St. Mary's Preparatory School.

Though our first teacher was a layman, Mr. Pongo, the preparatory school was run by more penguin people, this time Christian Brothers, mostly from Ireland. Looking back, I can see that they had their own reasons for getting access to our neurons: they were intent upon indoctrinating us with their perception of how to get to heaven.

I was a classical good kid. I swallowed it all, went to confession every week, and took communion at Mass on a Sunday. The brothers also had a more secular mission: to prepare us for the eleven-plus general intelligence test, the first of the hurdles by which future leaders would be chosen. The administrators of Her Majesty's Right Honorable Government would use the results of the exam to decide whether we were worth spending money on for a grammar school education, or whether society should avoid the ordeal of teaching us to think, and instead train us to work with our hands—in factories or trades.

In theory, the intelligence tests were supposed to be administered to kids without prior training. In practice, they included syllogism puzzles and knowledge of arithmetic tricks that allow rapid mental calculations. These can't be sprung on a naive child with the expectation of a coherent response—that had been the fantasy rationalization of a grown-up committee. It has all been changed now; I can only hope it has at last become truly rational. (But I tend to doubt it.)

What actually happened in my day was that those parents who understood the situation sent their children to prep schools, where they were coached in such tests, thereby raising their chances of getting the government scholarships. In effect, then, the tests functioned as a test of whether your parents understood the system and had the means to send you to prep school.

In another example of 'working the system,' the Christian Brothers Preparatory School delivered Catholic kids who could pass the government's tests and thus bring in scholarship money for the Brother's denominational grammar school. These were the circumstances that led to my small feet being placed on the ladder of upward mobility. The rung I eventually reached was that of a professor in the University of Maryland's Medical School, in far off Baltimore, which didn't have a whole lot to do with running England.

Poetry

My first exposure to poetry had an evident, specific purpose. The Liverpudlian version of the Queen's English is called the "scouse" dialect. Ideally the lips stay motionless, with the mouth just slightly open. Not only does this make it difficult to know where the sounds are coming from, but it also leaves the rapid-fire words somewhat difficult for the rest of the Queen's subjects to appreciate.

I think this is why we were early introduced to elocution, which involved acrobatic lip movements while reciting poetry. I learned the rebellious *Song of the Cornish Men* and found myself asked to "do" it regularly. The poem was about Michael Trelawney, a lord from Cornwall, who inspired great loyalty from his subjects. Because he made the mistake of pissing off King James in London, he was confined in the Tower. This had awful implications. When the sad news reached Cornwall, Trelawney's serfs gathered up pitchforks and rakes and began a long march to the capital city. The poem portrays their mood:

Trelawney's locked in keep and hold, Trelawney's days are few,
But shall Trelawney live, or shall Trelawney die?
Here's twenty thousand Cornish Men, will know the reason why!

King James decided to set Trelawney free before the marchers passed Bristol. When I think of Dr. James Hansen's attempts to warn us all about the impending dangers of climate instability and of the successful suppression of his well-qualified views by politicians on the hook to the fossil fuel industry, I fear we have lost something important since the Cornish men took to the road. I am, however, trying to save my rabble-rousing for later in the book, so enough of that.

When we were ten, we were taught by Brother Scaysbrick. He understood our sense of fairness, and for teachers, this was unique. Every parent notices the onset of this epiphany as small, powerless children come to comprehend that might is not supposed to make right in human affairs. Our enthusiastic affection for him was based on his evident understanding of the concept.

At the age of eleven, we were marched under the guillotine of the intelligence test I mentioned earlier.

Those who had not been well trained emerged from the exam with post-traumatic stress disorder, while the responsible authorities could gravely congratulate themselves on how 'fairly' they had sealed the youngster's fates. Because I passed the exam, I remained on the avenue to academia.

Epilogue
Cross-correlations

What did all that have to do with neuroscience? As we learn to comprehend the speech of others and then speak for ourselves, the left cerebral hemisphere is learning to break up reality into categorized details with verbal labels. From the words *I, me* and *mine* we develop a verbal sense of self. This becomes the left brain's most important object-detail. To repeat: consciousness of self develops in the talking hemisphere. Furthermore, by adding action labels—verbs—we become able to generate a semi-continuous approximation of what has happened to us: a personal adventure story or self-narrative.

I'll happily accept whatever sympathy you were willing to extend to my young self in the above chapter, as long as we don't slip into denial of the main point: the output of my brain was a series of loose associations, most of which were shallow and wrong: the duck pond wasn't bottomless; there was no witch in the boarded-up house, etc.

Why does this happen? The brain's first step in the thought process is to form loose associations. The next step should be a crosscheck against memory. I couldn't do that because these were original experiences with respect to which no memories were available.

For example, I only knew of one duck pond and I never did see it dried up with its bottom evident to the eye. Nor did I have a mental model of the way the world was laid out, and so I couldn't figure out where to go in it. As a consequence, my parents could tell me anything and there was little choice but to believe.

Intelligence

A contentious debate continues as to what IQ tests measure. In Richard Nisbett's book, *Intelligence and How to Get It* (2009), he points out that half of the population in 1917 had IQ scores that would be equivalent to 73 on today's scale—the mentally handicapped category. Moreover, the architects of a recent global economic collapse had undoubtedly passed the usual intelligence tests, but it turned out that their loose associations could not stand up to a crosscheck against financial reality.

Although children do get less gullible with age, and you will see me improve, albeit painfully, in the next chapter, there is a continuing serious question as to how accurate our thoughts can ultimately become.

Chapter 2
Emigrant Dreams

Fred

Frank

Albert

Wanted Men:
The Boyne Gang

During the sixteenth and seventeenth centuries, English frustration with Irish resistance movements led to draconian punishment: the land titles of the native Irish were given over to English landlords. Now tenant farmers in their own country, the Irish became impoverished. They subsisted on potatoes while growing plentiful supplies of wheat, barley, and corn—that were all fed to pigs destined for England as a cash crop, to benefit the new owners.

Although the land laws began to be repealed in the late 1700s, it was necessary to buy back one's ancestor's farm, and the people were by then too poor to do so. They generally continued to be tenant-sharecroppers.

When blight catastrophically destroyed the entire potato crop for successive years, absentee landlords, by then accustomed to a steady income, refused to allow diversion of the grain crops from fatted pigs. Between 1845 and 1849, the distant English Parliament debated how to restore humane rationality—while pigs thrived and peasants died.

A century earlier, the Irish writer Jonathan Swift (author of *Gulliver's Travels*) was so horrified by the living conditions of the Irish working classes and their children, that he wrote perhaps the most stinging satire about exploitation that the English language has ever seen. *A Modest Proposal* sets forth, with a straight face, a means to save Irish infants from the life of beggary and destitution that most could expect. Swift suggested that they should be boiled, fried, or baked, immediately after weaning, so as to incur no further expense. When a million Irish starved to death so that fatted pigs could go to England, that previously shocking Swiftian satire became mere irony (ref: Wittkowsky, 1943).

In response to the famine, another million, my ancestors among them, emigrated around the world. This is part of the deep history for which Tony Blair apologized to the Irish, when he became British Prime Minister in 1997. It was a good start to his efforts to resolve the Irish troubles of our own time. Queen Elizabeth herself recently visited Ireland. While there, she reiterated the wish that some of the past had never happened. I hope she meant the English arrogance. Some of her own ancestors, however, were killed in the troubles, so she was entitled to mean that, too. (I am trying to present, herein, a viewpoint that would have made it all unnecessary.)

In Liverpool in 1928, while Grandfather Andy Boyne finished the last two years necessary to vest his pension, his wife Mary took their remaining children to join grown siblings, who were then living in Sydney, Australia.

Shortly after the immigrants arrived, so did the Great Depression. They were hungry times. With no father figure, the trio of young boys, Fred, Frank and Albert, developed a gangster perspective and became hard to control. They stole food in the local markets, set fire to their school, and were expelled, after which they began to 'run away.' On one occasion, Mary went to the local railroad station to inquire if any of the "Boyne Gang" had been seen on the trains. She was walking home in the twilight when the Great Depression, for Mary, came to its close. The Grim Reaper's instrument was a Model 'T' Ford, being driven by a drunken driver; Mary died immediately. (I think this was one of many causal strands in Fred's later alcoholism.)

Fred and Frank then ran away in earnest, making it to the docks in Sydney. Fred signed a contract as a deck boy with the Moreton Bay, a freighter bound for London. His pay for the voyage was to be one shilling (twelve pennies). Upon docking, and asking for his shilling, he was told in tones reserved for Irish half-wits, "Oh, you don't actually get that!" With no union representative because there was no union, there was no way to sort out such a rationalization, so Fred had to borrow money from the mate to pay for bus fare to Liverpool to see his dad. Nevertheless, Fred stayed with Moreton Bay for several voyages and, in a few years, he made able seaman, but he never got beyond that.

The Tasmanian Devil

Frank took a different approach: he hid in the coal locker of a Liverpool-bound ship. Once the local pilot had guided the vessel out of Sydney Harbor, and returned to home, it was impractical for the ship to return to the quay. That was when the blackened stowaway emerged—to be put to work for the rest of the voyage. (Needless to say, he didn't get no shilling that way, either.) Unfortunately, Frank found that depression England was worse than Australia.

After a winter of sleeping outdoors and still 'finding' food at the local markets, Frank decided to return to the warmer climate, using the same cut-rate travel plan.

Although he also became an able seaman, Frank seems to have had considerably more self-confidence than Fred, and was more able to take advantage of opportunity.

With a lower center of gravity, Frank had focused his muscularity on wrestling. One day, when both brothers happened to be in Sydney together, handbills announced the impending arrival of a circus, and included a challenge to the locals to test themselves against the circus strongman—in a wrestling ring. Frank decided to fight. The photo may have been taken in preparation for the big night.

Of course I would like to report that he won, but he didn't. Nevertheless, it had been worthwhile: the professional realized that the youngster had an aptitude for being under the spotlights. Promising to teach all he knew, he invited Frank to tour Australia, taking on all comers, as part of the circus and Frank agreed.

Wrestling as *The Tasmanian Devil*, he progressed through the ranks and fought around the world.

At the peak of his career, he was offered a match against the world champion in his division, a Greek named Papadopoulus. The fight was to be held in the New Brighton Palace, just across the River Mersey from Bootle, where Frank had spent his early childhood. Although the result was a draw, and the champion's title remained with Papadopoulus, the older members of the family remembered the night as a high point of their lives.

Later on, but still in the pre-war years, Frank and Fred took advantage of their disparate appearances. They set up phony wrestling matches in Manchester, where no one knew they were brothers. This was when Fred used the name Gentleman Jim (which I would later apply to an invention). Knowing each other's capacities, the devil and the gentleman put on a good show and were well paid.

During the war, Frank seized more opportunities. He immigrated to America, joined the Army Transport Service, and trained Marines in hand-to-hand combat just before they set off to liberate the South Pacific Islands. After the war, he joined the Seafarer's International Union, this time putting his talents to work as the leader of a goon squad. (He battled with other unions for control of the manpower contracts in developing ports such as San Juan, Puerto Rico. The police played the role of referee.)

Fred Emigrates Again

In about 1950, Uncle Frank wrote to say that seamen's working conditions in America were far better than they were on English ships. This was thanks to Harry Bridges, a union organizer and another immigrant from Australia—a man whom the FBI had done their best to deport, presumably at the behest of shipping companies. Frank said that Fred could earn enough at sea on American ships to spend three months at home every year. Dad was enthusiastic, but Belle felt that it was all an excuse to ditch his family responsibilities.

His later explanation for leaving was that he couldn't earn enough to satisfy her expectations, either on the docks or on English ships. (This was a transparent piece of blame-game rationalization.) Sponsored by Frank, however, Fred became an American citizen. Back in Liverpool, we lived on an irregular flow of dollar bills that came in the mail. He knocked at the door, two years later, and, when I opened it, I didn't recognize him. When he asked if my mother was in, his voice stirred memory, I realized that I had a father again. Belle got to go on a retail spree, and everyone seemed happy.

The blissful period lasted less than a week, for Fred was eager to go down to the big pub at the docks, *The Gladstone*, to see his old buddies. He could do without alcohol for long periods, but seemed to be incapable of limiting his intake when out with the boys. He had stories to tell, and I suppose war reminiscences were high on the agenda. He always came home stinking of beer and whisky, and tucked himself into bed beside the wife that he now rarely saw. Invariably, he would soon awaken, needing to visit the bathroom.

Although we now had indoor plumbing, he could never remember which room it was in. Consequently, he might open the window and use the neighborhood, which impacted mother's status sensitivities severely. She would resort to the silent treatment, until he repacked and returned stateside. Their kissing and cuddling in the earlier years had evoked security and happiness in my bosom. But over time, their relationship only generated sadness.

Alcoholism was a common problem. I can understand why the men turned to self-medication, and why Belle hated it. I couldn't understand why she stayed so angry with Fred, for we continued to live on the dollar bills that came in the mail. Although I still regarded Fred fondly, the news that he was coming home eventually triggered expectations of emotional battles, and anticipatory nausea.

The Family Emigrates

Mother's favorite yardstick in life was Winnie McGarvie, a childhood friend and rival. During the post-war housing crisis, Winnie's husband, Jim, who was doing well working for the Cunard Shipping Company, sent Winnie and their kids to America for a six month holiday; he was turning out to be the kind of husband Belle would have liked. Winnie's letters from America made mother green with envy, probably intended. Around that time Fred had realized that America's Social Security System was likely to provide the best pension he could expect. Consequently when the notion that we should all emigrate sprang forth, it may have seemed like a countercoup to Belle, who agreed and announced as much to Winnie. We sailed across a stormy Atlantic in 1956.

Frank was about to take a two-year sabbatical on the strength of a quick profit in the 1950s stock market. He learned to use the aqualung, shark hunting in New Zealand. His main philosophy was to always attack, "Don't let them think you're scared!" Right. We arrived in Boston just in time to take over his car, a Nash Rambler.

We drove down the eastern seaboard looking for a place where Fred's union could provide work and where Mother could stand to live. Gas was cheap, but motel accommodations seemed expensive, so all five of us slept in the car on a couple of uncomfortable nights. I was small enough to curl up on the back window ledge. We first tried to settle into a house in Fort Myers on the Calahootchy River, in Florida, intending to either rent or buy it—if Fred could find a job. Since none came, we hit the road again two weeks later.

Something happened in Fort Myers, however, that is obviously relevant to neural circuitry. This subject matter rarely receives the frank public discussion that it deserves, so we are going to fix that. I should lead up to it with some earlier loose associations, as follows.

Mr. Pongo's Class

The toilets in the nunnery were all of the sit-upon kind, but the preparatory school for boys had standup facilities. As with the bluebells, memory may be a bit out of proportion: I remember porcelain arches cresting far above our heads and a rumor that they had originally been used by Roman soldiers.

With the innocence and creativity of eight-year-olds, we made the best of the situation; we had spontaneous contests to discover how high we could pee. This could be considered evidence of an early interest in physiology. We were so engrossed in competition, however, that we didn't notice the arrival of our teacher, Mr. Pongo. He saw the behavior as evidence that we were inherently evil.

A few weeks later, in drawing class, we were supposed to draw a man. I found that I had grown tired of stick figures and, with further innocence, tried to draw an anatomically correct man. This was sufficiently engrossing that once again, I did not notice Pongo, until his shadow fell across the paper. Apparently the man was supposed to have clothes on, and I was admonished to never do that again. (I don't know where Michelangelo went to school but it sure as hell wasn't St. Mary's.)

By age nine, my schoolmates and I were completely under the influence of the Christian Brothers. Old Brother Cummings was particularly mysterious. He would occasionally draw me aside on the playground and ask what I called that thing between my legs.

"It's a dick, isn't it?" the tall black entity would say, and I'd reply, "Yes, Brother," and escape to play, aware of an aura of something inexplicably creepy, and wishing that the grownups would make up their minds about these things. Prurience was certainly being manifest, but not in our minds.

Pilgrimage to Lourdes

The brethren frequently admonished us never to get in bed with a woman, even your mother, and never to look at two dogs together in the street. My maternal relationship didn't involve early morning snuggles, so that prior restraint was irrelevant.

It was the two-dog thing that really puzzled me. Knowing that you weren't supposed to look made it inevitable that you would. Apart from butt sniffing—which just seemed really weird—I saw nothing to explain the admonition. But Tinker would sometimes disappear for days on end. I never found out what he was up to.

A potential clue came during my first year at the grammar school, when my class went on a pilgrimage to Lourdes, which is near the border between France and Spain. We sailed from Dover to Calais, boarded a train, and stopped overnight in Paris. The class prefect, Nelson, had been assigned to keep order amongst us. This foresight left Brother Patrick free to bow his head and clasp his hands at his robed waist, as he led the way through the streets of Paris—a model of religious decorum. In a supreme irony, however, our travel plan required that we cross the street in front of the Folies Bergère Theatre.

A huge poster near the entrance portrayed a lovely young woman, nearly wearing an off-the-shoulder blouse, one side of which was off the breast. When Brother Patrick, presumably for the first time in many years, found himself eyeball to nipple, so to speak, he stopped in his tracks. This caused a pileup in the line of schoolboys behind him, hardly Nelson's fault.

As we recognized the novelty of the image, our mouths fell collectively open. My brain began wondering how on earth I might get past the ticket collector, and go to see the Folies.

This is a normal biological response, and all but one of us seemed to feel the same way. Brother Pat was the exception. His black robed body curled up as though a rugby ball had emerged from a rent in the space-time fabric and landed right in his belly. To the accompaniment of honking car horns, he charged across the road, evidently looking for the corresponding goal posts.

I suppose he thought we were right behind him. That we were not can be blamed on Mickey Butterworth. You see, for weeks now we had been regaled with stories of the miracle in which the Virgin Mary appeared in Lourdes to little children—just like we were.

In response, the senior Butterworth, evidently a holy man, had considered the possibility that Michael might get lucky and see her too. So he loaned him the family camera. Quickly responding to the original shock, Mickey decided that the moment had arrived.

He lifted the Brownie and squinted through the viewfinder. The rest of us realized that there was indeed a God in heaven.

With one shot taken, Mick looked around and we all grinned at each other. To be on the safe side, someone told him to take another, and so he was still winding the film forward when a black-robed dervish, whose face was by now purple, blasted out of another rent in the fabric. Michael was grabbed by the ear—the brothers always grabbed us by the ear in times of dire necessity—and we were led firmly away from this famous den of iniquity. Fortunately, the camera was on a strap round Mickey's neck, so it didn't get lost.

Later that evening, however, we learned that staring at exposed breasts was even worse than looking at two dogs in the street, and might well have resulted in the loss of our immortal souls. I thought I must be pretty thick because I still didn't get the connection. (I was also preoccupied with trying to figure out what favors I could do for Mickey in the days ahead.)

The frontal lobes of human cerebral cortices are a recent evolutionary advance. They are thought to permit and support foresight, and they continue to develop well into late adolescence.

Eleven-year old frontal lobes, however, are not yet able to see very far ahead. Had ours been more mature, one of us, me for instance, might have foreseen what would happen next, and safely remove the roll of film from the camera. As events unfolded, it was Brother Pat's forebrain that did its work during the night, and that holy man awoke in the morning with the associated benefits. At breakfast in the hotel dining room, he asked Mick to hand over the camera.

Holding it in both hands, he studied the film counter. With his suspicions apparently confirmed, he surveyed our faces and then, very deliberately, opened the back to remove and unroll the film in the morning light. We were just smart enough to try to prevent our inner horror from being too obvious. Had Mickey been able to bring home the goods in 1955, he could have become the richest little schoolboy in England, but it was not to be.

At about the same time, thousands of miles away in America, an adult version of Mickey Butterworth was also photographing uncovered breasts. Marilyn Monroe had even consented to stretch out on satin sheets, in all her nippled glory, and Hugh Hefner made her the centerfold of an early edition of *Playboy*, thus sharing the pleasure widely and profitably. Half a century later, topless lady images in England are as ubiquitous as the daily newspaper. In my childhood, however, adults were still being jailed for circulating what were called 'filthy' pictures.

So, if we had in fact brought home the booty, so to speak, this tale could have turned into a full-scale tragedy. Consequently, I must admit that Brother Patrick probably made the right call, albeit in response to a then prevalent set of perverse rationalizations.

Ft. Myers

You may have forgotten, but I was in the middle of telling you about our immigration to America, when we were diverted to Paris. Before we return to the main line of the narrative, I need to fill in my only other piece of sexual knowledge; it won't take long. Whenever Fred was in England, sitting with us watching television, and a bra ad came on, he would say, "Hmm, she's got nice shoulder blades!" That was it. Now back to Fort Myers.

At the back of the wooden garage attached to the house, a ladder was built into the wall. It led to the open rafters.

There was no evident reason to go up the ladder, but I had nothing else to do, so why not? Once my nose was level with the top plates, I was surprised to find a glassine envelope full of photographs, lying there in the dusty darkness. I hooked my arms around the rails, freeing my hands, and found that the topmost print was of a rather pretty fair-haired woman, reminiscent of the miraculous apparition in Paris.

Although this lady's sweater was pleasingly snug, it was completely on and revealed nothing beyond those interesting contours. In the next photograph, the girl's hands had crossed and dropped to her waist, where she had grasped the lower margin of that sweater. She gazed into the camera and, as I gazed back, I could almost hear gears begin to shift in my mental mechanisms.

I proceeded incredulously to the next print. Now the sweater was nowhere to be seen and a highly detailed bra-ad was in my personal hands: I could see the curve of her so-called shoulder blades quite clearly. They looked gorgeous. Said hands began to sweat.

Did I mention the heat? Florida is in the semitropical latitudes. If you climb to the top of a ladder inside a wooden garage, you arrive in the tropical latitudes.

In the next view, the lady's arms had shifted to behind her back. A heritage of prudish repression left me incredulous but my forebrain was well enough along that I saw what this implied for the many remaining photographs. The presumption drove some sort of neurosecretory squirts that made my head swim. Fearing a rung-free descent to the concrete below, I dropped the packet and gripped the ladder rails before crawling down to the lower latitudes, heart pounding.

Did all the years of exhortations that attempted preemption prevent me going back up the ladder to stare admiringly at a pretty woman in the buff? No, they did not.

Moreover, I think formal sex education in schools is an excellent idea. With pictures. Adults who can't get the rationality here should probably not have children.

Jacksonville, Florida

England's early botanic grandeur of oak forests had been turned into man-o-war sailing ships long before my time. Large beasts and reptilian critters went the way of the forests; manicured gardens and hedgerow prettiness are all that remain. When Fred finally found a job, however, it was on the Saint John's River at the docks for the Sealand Container Corporation in Jacksonville, Florida. This geography would lead to considerable critter exposure.

We arranged to buy a brick bungalow at the end of Pecan Street, in Arlington. As my brother and I spilled out of the Rambler on the night of our arrival, we realized that the new house was at the end of a road that was still being pushed into the swampland. Furthermore, there were anole lizards all over the lawn! We immediately began the important business of stuffing them into a cigar box.

The day our immigration began to fall apart may have been triggered by Belle's discovery that the cockroaches in Florida are humungous, and they levitate. (When she first opened the front door to our new home, and a buzzing creature made a flying escape, however, Fred said it was a humming bird. This delayed the inevitable reaction.) When brother and I came inside and inspected the rooms, we found that each window had its own lizard patrol, evidently competing with big hairy spiders for a harvest of flies. This caused another shift in mental gears, and a sudden doubt registered in consciousness.

Brother and I cracked open the cigar box just a little, to behold a scaly version of the black hole of Calcutta. Then we looked at the free-range lizards... and then considered the box again.

We were experiencing a shift in context understanding. Once enough reconsideration had taken place, all the fun seemed to drain away from the lizard-catching business. I hope they all found their way back home.

I donned a swimsuit every day, took a bow and arrows, and headed out into the Florida jungle.

The wildlife was plentiful, and much of it was dangerous: water moccasins, rattlesnakes, and huge turtles with sharp jaws. I was in heaven. The folks didn't figure out school requirements until just before we returned to England, but I got to ride the big yellow bus for three weeks and was thrown in with adolescent Southern belles already into lipstick and makeup. There was no denying how pretty they seemed; there was no stopping thinking about them; they were even better than lizards.

A Penny Drops — Learning to Learn

The house foundations began to crack six months after we moved in, and our lovely modern bungalow began to sink into the swamp. Fred was mortified; he had screwed up again. Belle decided that we were headed into the economics of white trashhood, and must reverse the emigration. It is only plain to me in retrospect that we had been thoroughly impoverished sons of the sod all along. We packed up, drove to New York, and sailed for England.

Back in St. Mary's, with my mind still in Florida, and with eight months of the school year gone, I found the course content incomprehensible. The thought of catching up in the liberal arts was laden with an aura of impossibility. Foreign languages and geography bored me. Although history was full of good stories, I had missed that year's allotment and couldn't see the big picture. When the final exams were graded, I was declared last in class.

Deep immersion in the family's experience of failure, however, had given me a personal view of civilization's canvas.

I understood that Fred was unable to accumulate enough money to effectively control his and our destiny, and I had sensed his depression at the scope of this failure. I responded with a desire to earn enough to guide my own fate more deliberately.

In my inner dialogue, a verbal mantra developed: I would strive for a "better job" by doing well in school. An ancient Greek might claim that his breath began to move faster; a romanticist might say that a burning yearning filled his heart; but I now describe it this way: my forebrain was really kicking in. Although I had never heard of Thomas Paine and personal sovereignty, I had begun to sense the possibility and so to develop the willingness to reach for it.

Fortunately, science teaching began the following year, and I finally found myself in long trousers, starting science along with everyone else. Physics came up first. A brand new lay teacher, whom I remember fondly, described electromagnetism and told us how the theory had been turned into the practicality of the electric bell. (I loved this transition between theoretical thought and practical application.) He encouraged us to copy the bell design into a notebook. I went one step further, using pocket money to buy one of those etch-a-sketch boards based on a magnetic trick by which you can draw and erase over and over without using reams of paper. It was a bit Neanderthal to be called the digital tablet of its time, but it was a great investment.

My study was the coat closet behind the entrance door of our tiny flat. Learning the circuit took much repetition, but learning-by-doing was effective. Before going to bed that night, I had an impression that the electric bell had shifted from being an external two-dimensional drawing to being a multidimensional something in my mind. I could now start from any point in the circuit and either work or remember my way forward. I was relieved to find that I still 'knew' it in this manner when I woke up.

At the next physics class, our teacher began by asking if anyone could draw the bell circuit on the blackboard. In my mind's eye, I can still see where I was sitting, apparently remembered because the moment coincided with intense emotion.

You may expect to learn that I threw my hand up right away—to establish that I wasn't as dumb as it might have seemed. But that common desire had never been a particularly influential part of my mental makeup. In fact, the certainty that I had learned something the night before was giving me all the ego satisfaction I needed. Moreover, as I later learned, the attempt to be impressive has a habit of negating the achievement—now there's a convoluted thought!

In any case, my forebrain had raced into Machiavelli mode, and my inner dialogue had suggested that by keeping my hand down, I might get a perspective on what this studying business had gained me, in comparison to my classmates. That did seem valuable. With no volunteer, our teacher encouraged a known nerd to the front of the room. I watched the nuances of his body language intently. His evident befuddlement at the blackboard left me in a state of controlled euphoria. I understood that I could absorb the knowledge that my teachers were offering; that I would never again come last in class; that the result would be power over destiny. I felt deeply thrilled.

I imagine that all teachers hope to infuse such motivations in their students. Unfortunately, it is hopeless to attempt to do so by verbal exhortation. In other words, there is a big gulf between telling and teaching. My genuine motivation had come from that vicarious and still painful experience of my father's dilemma.

The closest thing a teacher can provide is a relevant, heartfelt story. Stories require the teacher to provide cues for hemispheric switches, which seems to drive students through the hemispheric tick-tocks that are needed to cross-stitch information into a durable mental fabric.

In contrast, admonitory exhortations arising exclusively from the teacher's status-obsessed (and/or sex obsessed) speech centers travel with equally selectivity into the student's left hemisphere, and leak out overnight, if not on the way home.

The other science subjects also fascinated me. Miss Shehan taught biology. I listened and absorbed details of animals and plants, relating it to my solitary wanderings through Florida's forests of trees draped with Spanish moss—and fell in love with her.

Chemistry came from one Brother Coleman, inevitably nicknamed "Dusty." (Working with coal leaves one covered in black coal dust. He wasn't actually a coal miner … oh heck, perhaps you had to be there.)

Dusty Coleman explained chemistry well enough, and I got to perceive the patterns, but I did not like the man. I remember his parting words of wisdom, when we eventually graduated, his valedictory: "Stick to beer as long as you can; the whiskey will do you in." This was a further hint that the civilization we were being prepared for had some unresolved problems.

Fred's New Ambition

Fred shared the Irish susceptibility to deep wonderment at technology, and his sea bag often housed American stuff that seemed magical to our Liverpudlian minds.

On one occasion, the mystery sack revealed a machine bearing two reels of brown tape. After switching the American plug for an English one, he inserted it into a wall socket and the reels began to turn while a humorous ditty was being played on the radio. When the song finished, Fred switched off the radio, rewound the reels and then replayed, *"Granny's Old Armchair."* It was a tape recorder.

A gypsy had once told Dad that he would die rich. He now imagined that a tape recorder business would do the trick.

The recorder was certainly a big hit with me; I immediately replayed the song till I had written out the lyrics. Then we smelled burning, followed by black smoke. Several details had overflowed Dad's bucket.

English electricity is driven with the force of 240 volts, twice as much as in the US. Different electrical plugs on appliances in different countries are meant to prevent them being plugged into the wrong source, for too much voltage drives too much current, and overheating melts the insulation around the motor windings, and so on. I further suspect that the tape reorder was a Grundig, already being made in Germany, so that Fred's broad nascent notion of bringing them back across the Atlantic was rumpus reversus from its inception.

A few days later, Fred confided to me a story of his mother, Mary. She had come from the peat bogs of Ireland to be a scullery maid in a well-to-do Liverpudlian household. Early in her service, she was asked to prepare some French beans for the Sunday dinner.

Although unfamiliar with such beans, she noticed their resemblance to peas, and carefully cracked open the midget pods, filling an eggcup with the tiny harvest. The pods were thrown away. When the eggcup was delivered to the table, the guests ended up rolling around on the floor, laughing till they cried. This seems such a harmless little story as I look back on the words. One has to do a recursion to see deeper.

The country wisdom of the Irish was out of context in more technologically advanced societies, and this fed into the human love of ridicule. When Fred shared the tale of Mary's mistake, his inherited embarrassment was palpable. His crestfallen depression, as the tape recorder had gone up in smoke, had been part of a lengthy tapestry. I think those feelings further empowered the monkey on his back, and urged the resort to whiskey that eventually did him in.

The Boxer

I didn't know that I was enjoying a more sheltered life than Fred and Frank had experienced, and so I made the mistake of assuming that, if they were unbeatable, then so was I. This photograph seems to betray that assumption. It was probably the first of my own rationalizations that was thoroughly dismantled by reality.

When I was about eight or nine, I had squared off against the son of a dustman on the sidewalk near our houses. I have a definite recollection of the grogginess with which one punch rendered him the victory. As often happens with dogs and men, we then became friends, and he generously invited me to come into his house and watch some cartoons.

In the days before we British had television, I couldn't imagine how this would be possible. It turned out that he had received a small film projector for Christmas. The screening room was to be the whitewashed brick wall inside the outdoor lavatory.

The standard toilet design of the day included a tank full of water near the ceiling. A pull chain activated the plumbing so that water descended through a fat pipe with enough force to give the sit-upon unit a good flush.

Consequently, the cartoon image, on its way to the brick surface, had to wrap around the pipe and chain.

(I still feel spoiled rotten when I enjoy today's high definition images while sitting in an easy chair and smelling nothing in particular.)

But what had I learned about aggression? Apparently not much, for a second descent into violence came a year or so later, while pushing a bicycle across the local playing fields. An older boy, who was following along, began to casually kick the rear wheel with the imperial insouciance of someone bigger and stronger. After my verbal objection, the classic street-challenge was uttered: "What are you going to do about it?"

I was going to think, is what. And now I'm going to get a little wonky about what that thinking involved.

At the time of all these experiences, of course, I didn't even know I had cerebral hemispheres. Yet the episodes were somehow flagged as mental treasures, while great stretches of the surrounding weeks and months became blank. How, why? We have learned a great deal about the how in the last fifty years, and a key figure in that research has been the Nobel Laureate, Eric Kandel.

Perhaps driven by experiences on Kristallnacht in Austria during 1939, when German soldiers commandeered his family's possessions—including a toy that was precious to a small boy—Kandel became a psychiatrist. But he soon despaired of the rate at which psychiatry was learning to comprehend the human condition, and turned instead to the marine snail, Aplysia.

Kandel hoped that he might learn the mechanism of memory in a creature that had only a few neurons and that could remember very little. Academics love this kind of counter intuitive flourish. As in many other cases, resort to a simpler animal system was crucial in winning the basic insights.

Kandel and his associates discovered that experiences worth learning drive long-lasting variations in synaptic efficiency.

They also established that mammals modify their synapses in the same manner as snails (see *In Search of Memory*, Kandel, 2008). So... we have substantial insight into how a memory is written.

It turns out, however, that we don't remember everything that ever happened to us. Instead, we select certain memories and mark them for remembering. Diane Ackerman (2008) has nicely articulated the current standard explanation: an emotional acid etches dramatic memory permanently into place.

My backlog of memories certainly conforms to the emotional acid metaphor, but there is another consistency that is worth separate attention. My remembered stories are of situations wherein the unfolding of reality forced reexaminations of the context. In other words, they appear to have been instances when a hemispheric tick-tock was useful. Going further, I have the impression that the 'best' stories involve multiple tick-tocks. Now let's go back to the playing field and the bullyboy, and watch the differing context comprehensions in slow motion.

The challenge—to fight a larger boy—was obviously emotionally arousing. Instead of reacting blindly, however, I am struck by how deliberately I reviewed the overall circumstances. The first thought that arose was this: the heavy, old-fashioned Raleigh bicycle in my hands was an intrinsic asset. Since I hadn't flung myself at my provocateur, but already knew that I was going to, I also had a quite delicious, even less obvious advantage: timing. To a third person observer, it would still seem that a large boy was dominating a smaller. To a mind scanner, however, my brain patterns might have matched those of Sir Francis Drake as he rolled bowls across the green at Plymouth Sound, while the supposedly invincible but lumbering ships of the Spanish Armada hove into view. I had never performed the intended choreography before, but was confident that I could string all the little component actions into the necessary maneuver.

Neuroscientists call the small movements: 'fixed action patterns,' or FAPs. Our brains allow us to string FAPs together to create new sequences as fast as we think of them—as in a football match, or free form dancing. I decided that the next kick would serve as the trigger. When it came, I lifted the bike into the air, and then began to move it in an arc, which meant stepping into a circle and compensating with a crouch as it began to revolve faster. The process of imparting the necessary momentum was relatively slow. In the initial phase, however, I was moving it away from the eventual target, who may have been simply puzzled.

He was presumably less puzzled but was still frozen in place when the bicycle had become an unstoppable force. Soon flat on his back, he no longer needed to wonder what I was going to do. Being still conscious and relatively unharmed, however, he quickly disentangled from the bicycle. With no Plan B, I was again beaten up. But there was something token about it; my dominator suddenly seemed to be in a hurry to be getting on home.

Let's see... that's two defeats. Had I learned yet? No, I had not. The third time I fought, there was no doubt about how soundly my nose broke. My good friend Terence objected to a call in a cricket match and decided to go home. This would have been perfectly acceptable, but it was his cricket ball, and he planned to take that with him. After I objected, bare knuckles boxing stances were adopted. He was taller and his reach was longer than mine. I seemed to be a natural counter-puncher and this time I failed to perform the necessary context revision: smart short guys get in close where reach is irrelevant. I wasn't that smart and paid with my nose.

Said proboscis bled profusely, which I'm sure shocked Terence (who later became a priest), and the fight St. Petered out.

If you are still counting, that's zero for three.

Fred may have decided to do some real parenting at that point; he turned one of his sea bags into a punching bag with a permanent place in the back yard. I spent hours winning world championships against a passive opponent.

And so it was that when Mr. Highton, a new gym teacher, turned up at St. Mary's, I was still in denial. We were about fifteen. He started a class by asking for volunteers to don boxing gloves. This being the family business, my hand was in the air right away, and Pete and I were gloved up.

This time I was the taller, but Pete was stockier and, unfortunately, seemed to know all about being aggressive—I was soon receiving full-face smacks from his gloves, and backing up all around the ring.

My earlier trials and failures had not been completely in vain, however; they had left some residual benefit. First, I quickly realized that the sensation of a soft glove landing on the face was much less authoritative than a fist; it left no groggy disturbance of mentation; it drew no blood; it wasn't that big a deal. On the other hand, the softening effect seemed to operate in both directions, and despite my countering blow for blow, Pete, a diminutive Rocky Marciano, just kept on coming. That wascally sea bag had not prepared me for this.

My hemispheres put their heads together, so to speak, and I perceived a larger picture—an inherent vulnerability created by Pete's aggressive frame of mind.

After the next face punch, I feigned alarm by accelerating my retreat, but I was now in hidden charge. As Pete surged forward, pulling back his fist with great enthusiasm, my eye caught his sudden shock as he registered that I had bounced on my rear foot and that my approaching fist was well ahead of any defense. The blow canceled all forward momentum; he was finally at a standstill, fists dangling at his sides.

My inner dialogue delivered the thought of *going in for the kill*, (I must have heard the phrase somewhere) but the emotional state portrayed on his face had delivered another epiphany; the idea of hitting someone lost its charm.

Mr. Highton seemed to understand what had happened, and

called an end to the proceedings. Later I learned that most of my classmates thought I lost "because I was backing up the whole time," which must have hurt my ego enough for me to remember it. In a follow-up text, we will consider the hypnotic value of one-liner mentation, of which I think this was an early example. Nevertheless, I could see how every moment but the last had left the impression that I was getting whipped. But I didn't feel any need to argue; from inside the experience, I knew that I had finally won one. It would turn out to be the last time I would have a physical contest.

Pete died in a motorcycle crash about a year later, but he lived on in my mind, for the political battles of professional life always seemed to be a rematch. Furthermore, my hard-earned but now visceral understanding of context assumptions and timing made me, as we will see, a more formidable opponent.

Epilogue

In the epilogue to Chapter 1, I noted that the verbal hemisphere categorizes reality into those details that we label with words. The most important of these is the self. But each detail is embedded in a complementary context, and this latter domain is the fascination of the right hemisphere. Consequently, the view from the right is of the self embedded in the context and, as the scale gets larger, the representation of the self becomes reduced. Thus we feel small in a cathedral, or when we look out at the Milky Way. But we always loom large in the self-consciousness of the left hemisphere.

Splendidly practical as this evolved strategy has been, its pragmatism obscures a fundamental vulnerability.

Reality is not actually divided up into clear categorical objects. Instead, reality is a continuously interacting web of matter and energy. In other words, to the extent that verbal language gives us an impression of sharp distinctions, it betrays us. Fortunately, the broader wisdom of the right side, if solicited, can come to the rescue. Unfortunately, said solicitation is not obligatory.

As an example, consider Xeno's paradox. The logical side of the paradox asserts that, in order to go from A to B, we must first pass a point that is halfway between them. Once there, however, there is another halfway point for the remainder of the journey. Since there is an infinity of successive halfway points, logic seems to insist that we can never arrive at B. The paradox arises because we know that we can move from A to B; that a hare can overtake a tortoise.

Alleged explanations of the contradiction have focused on the question of whether a line can actually be divided into an infinite number of points. For me, the contradiction is better resolved with the understanding that 'half-way' to somewhere is an invented verbal concept, one with no actual properties in external reality. Consequently, a hare experiences no resistance as it crosses an infinity of halfway points. All such verbal logic, even that claimed to be 'bullet-proof', stands on similarly shaky grounds.

Bottom Line: The categorical verbal 'logic' of the verbal hemisphere is frequently in error. Nevertheless, the self-centered view of reality has had survival value, or it would not have evolved. What then, are its strengths and weaknesses?

Verbal logic seems well adapted to recognizing immediate cause-and-effect threats to bodily survival and to seizing short-term advantages. Where it falls down is in developing long-term rationality and ensuring long-term survival. For these purposes, something better is evidently essential, such as broader comprehension of the unfolding of slower contextual consequences.

The experiences just recounted in Chapter 2 included semi-conscious guidance from previous memories (e.g., fists are hard; these boxing gloves are pretty wussy). Furthermore, I seemed able to be in two places at once: I could either peer fearfully through the limiting gun slit of self-consciousness, or I could call upon a detached eye-in-the-sky perspective. The latter had proven capable of disclosing advantages that had not been apparent in the more usual view.

I seem to have thereby discovered for myself the essence of martial arts: transcendence of natural fear-for-the-self so as to have access to the context comprehension of the non-self hemisphere.

Chapter 3
The Working World

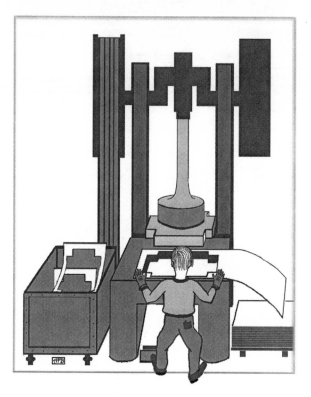

Factory Life

I had spent my childhood summers wandering the banks of the River Mersey, or the trails in Sniggery Woods. At the end of one school year, I was surprised to discover that we had passed beyond the limits of the child labor laws. We were fifteen and, instead of playing for the summer, we could hire ourselves out.

The gang at school all signed on at a tin-box plant. As the regular workers took their vacations, we schoolboys were employed to sit in their places for a week at a time, operating massive machinery. I started at a cast-iron punch press, feeding half a sheet of metal into its maw.

When I pressed my foot down, a clutch engaged and a Maltese-cross shape was whumped out of the metal. Then I would roll the back half into place, insert, clutch, and whump. A moment's delicacy was required to toss flimsy residuals into the scrap bin, before I turned to push and clutch, whump and roll, and carry on.

The sides of the boxes were folded up in the next station; the corners were soldered in a third; and the standard container for Oxo bouillon cubes came out at the end of the line.

While there was novelty in the first day of working at any particular station, piecework quickly becomes boring. Fortunately, those of us who had passed the pseudo-intelligence tests, and so made it into grammar school, could expect to find a less mind numbing form of eventual career, but the factory fate that awaited other children shocked me deeply.

There were other surprises.

We were now exposed to the secular world that we had been warned about for years. During tea breaks, older and bolder girls teased me gently. To my embarrassment, one noted that I was a handsome chap; that I must have taken bras off lots of girlfriends. Only much later did it occur to me that an admission of innocence might have had its reward.

I did develop a crush on a slender, delightfully pretty blonde confection but kept my distance. I always think of her when Mick Jagger sings *Factory Girl.*

Returning to school in the fall, more focused than ever on the notion of a better job, I spent my evenings filling notebooks with careful records of the chemistry, physics, and biology that I had scribbled down in class. As the next summer loomed, we again looked to the factories of our Merseyside world. This time, I ended up in a pea cannery.

Are yuh glad?

A curious bit of adolescent repartee began its spread that summer. Whenever enthusiasm was expressed, as in, "Hey, it's payday, today!" someone would ask, in those well-controlled, lipless tones, "Are yuh glad?" The question inserts a spanner into the mind; one feels embarrassed for having momentarily dropped dispassionate English propriety (or teenage cool, I'm not sure which). The mind jams; one finds no basis for a reply. To say, "Yes, I'm glad because I like to have money to spend ..." is like the unfunny explanation of a joke. In the game of verbal swordsmanship, a well delivered "Are yuh glad?" was a mortal thrust.

Pea pods ripening relentlessly in the fields had been pressured during harvesting so that they would split, thereby spilling the peas—but they still needed to be separated from the empty pods. The mess of plant material had been collected in twenty-eight pound aluminum trays, stacked on a flatbed lorry and taken to the courtyard of the cannery. Our job was to tip the trays onto a hip-high shaker—a flat plate inclined at about twenty degrees and full of pea-size holes.

It jigged slowly, back and forth leading the unwanted greenery to slide down the upper tray, while peas found holes, and so dropped through to the next solid shelf. From there, a jet of water swept them into clear plastic pipes, and they became a slurry of green shot, whizzing around the factory.

The other boys began making suggestions about girls they would like to get on the shaker with. I figured they were joking.

The foreman, a likable chap, organized a staggered departure for our meals so that the shakers were constantly being fed. On the first day's lunch break, I was sent off with Joey, one of the deselected who had been ejected from the educational system a year earlier.

Going to sea was the major alternative to factory work, and Joey had sailed for New Zealand during the European winter. The highlight of his voyage provided the lunchtime entertainment: a tale of rendezvous with a Maori lady of the evening. I said little but my face must have bespoken fascination that kids my age were already doing it. The story replayed in my mind that evening, and I slept at the foot of old glory (a reference to the staff that stiffens with arousal; nod to John Prine).

The next day, the foreman sent me to lunch with Joey's friend Billy, who had also gone to sea, but his first voyage was to Africa. To my further amazement, I was again regaled with a tale of sexual awakening, this time with an African lady in a grass hut. It proved rather less romantic, for she had been chewing gum and reading a comic book through the whole encounter.

The climax came the next day, when the forces of chaos resulted in my having lunch with Joey and Billy together. As we trudged to the canteen, making small talk, it struck me that they would have told their tales to each other already, and each had now told me. It was eminently clear whose turn it was to tell a tale, and there was no doubt what the subject matter was supposed to be. I stared into the void: I hadn't even kissed anybody properly. We collected our tea and took our lunch packets to a table, all this amidst a pointed silence from Joey and Billy. And me? I was longing for the Luftwaffe to come back and finish us off.

Suddenly, my brain retrieved a detail that resonated with my new friend's fascinations. I followed the ascent of his teacup, waited until his eyes met mine, and then said, "Billy, do you know how worms fuck?" The cup paused. Curiosity expressed itself in a lifted eyebrow. The tea returned, untouched, to the table. His head tilted, an eye narrowed, and he said, "No…how?" The hook was in. Now to set it. I spoke carefully, for instant comprehension.

"Every worm has both a female sex organ and a male sex organ, at opposite ends of its length." His eyes dilated and facial strain confirmed that a keystone in a mental archway had begun to wobble. With thespian incredulity, and emotional disgust, he said, "So...Do they fuck themselves??!!"

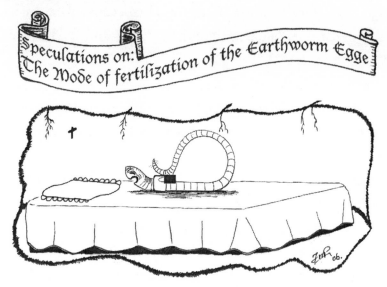

I was full of pride as I explained, with calm authority, that the real behavior was still more marvelous. Two worms would meet, negotiate the intention, and align so that one ran in the opposite direction to the other. Each could then do unto, and receive simultaneously from, the other. To finish with a flourish, I suggested that it might be twice as good for them as it is for us! (As if I might know.)

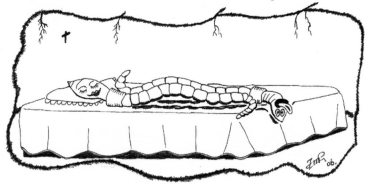

For Billy, this alternative to human sexuality seemed to be a wonderful insight, and this left me feeling high on my first teaching experience. Joey, watching from the side, decided that my ego needed adjustment. He said: "Are yuh glad?" Allowing silent seconds to tick by, he followed up, "Have you never had any, then?" There was no way out; I admitted to virginity.

Nevertheless, I had held up my end of conversational convention, and now we were all on the same psychological shelf, sharing important life issues. They volunteered the services of a friend of theirs, a lady who would do it with anybody in a back alley. Adjacent rows of our houses were built back-to-back, and a narrow alleyway separated their garden walls. It was in this Garden of Eden that Joey and Billy proposed that I taste the fruit of the tree of sexual knowledge. I did not accept their kind offer. Instead, when all the peas were canned, I retreated to the social shelter of life at a parochial grammar school. I had many years of snogging on the way to Sniggery Woods, in the woods, and coming back again before I finally did it, and I wouldn't have missed all that for anything that came later.

Career Planning for the Kids

Fred sometimes took me for precious-in-memory walks around the Gladstone docks, his old stomping grounds. On one of these, when I was probably sixteen, he asked if I had thought about what I wanted to be when I grew up. He suggested that radar operator on a ship might be a worthwhile goal.

In a seaport town, many looked to the sea for a livelihood, and I supposed that he saw radar operator as a large enough ambition. I thought he must have envied the guys who never had to go out on deck in bad weather. It simply didn't dawn on me that his larger frame of reference included a completely different perspective.

Decades later, I realized that he had seen all of his children as a tremendous burden, and had wanted us to be out of the house as soon as possible. While he didn't force me to realize how he felt, he did force this understanding on my young sister, some years later. When she was told to get a job, any job, in her mid-teens, she was considerably traumatized.

At the end of that school year, another round of exams came our way, another sifting. Coming up short put one out in the workforce. This must have been what Fred was hoping would happen.

I passed, however, and still did not notice that he was horrified that I would continue to be an economic burden. Yet, as I remember, we had lunch in school every day as part of the resulting scholarship, paid for by the government. It wasn't much of a burden. Fred's real problem was that his reference system, his context understandings, had developed during the Great Depression and the following War, and they had not been updated.

Career Planning by the Kids

Children in England associate the seashore with holidays. When they learn that you can get paid for playing around in rock pools, many decide to become marine biologists—and so did I. But in an overcrowded society, where good job opportunities required the long wait to fill a dead man's shoes, it was obvious that the competition would be merciless, and ultimate failure a likely result. In the first year of sixth form, I decided that love of biology had to be combined with something that would winnow the competition.

When I heard of a specialty called biochemistry, it seemed ideal. Yet how can one be sure?

The answer here is simply that one can't, at least not in the current chaos that we call a first world civilization. Nevertheless, the mysterious reality field that surrounds us often rewards decisive behavior; there is some value in getting decisions made, right or wrong. After all, trial and error really works.

After finishing the last two grammar school years, and still more testing, I was accepted into Liverpool University. The costs were again born by the government, and I was on my way to that nebulous better job. At that point, Fred was buying a substantial house in Cambridge Road, near the duck pond, and I was doing repairs and maintenance, experiences that I would eventually turn into the ability to build for myself. Fred was finally torn between pride and uncertainty about whether I would ever leave home.

My younger brother would soon bail out of the family disaster, first becoming a carnival barker and art student, and then a museum apprentice. He learned taxidermy while living under a staircase in a distant town. My sister eventually became a medical secretary and then found a niche in the computer industry, helping others over the associated learning barriers.

Epilogue

When Fred retired, coming home from the sea for the last time, he shifted from being an episodic alcoholic and became tipsy every day. Belle soon filed for divorce. It was granted, but the settlement phase revealed that their home would be sold, and its cash value split between them. Neither could then have afforded to buy a house. A compromise was reached in which Fred did life in the attic, literally living, sleeping, and dying up there. Belle ran the household. Many years later I came back as they approached eighty.

I was still trying to understand it all. As I walked along the sidewalk in the local village with Fred, I asked him directly why he had married Belle. The answer was just as direct. "For sex." The obvious follow-up question came to mind. "Why did you have me?"

With a similar lack of hesitation, he said, "Belle hated working in a munitions factory during the war, and there was only one way out of it." I was stunned and stopped to look into his eyes. He was dead serious. I got an impression that he had been dying to tell me this forty years before.

I suddenly saw our relationship in a completely different light, and began to laugh. (Sudden enlargements of perspective have a way of provoking laughter, even when they are fundamentally tragic. I will later argue that laughter is driven from the usually silent but context-appreciative right hemisphere.)

I realize now that my eventual father, the youngest of the Boyne gang had seen too much poverty in his life, and that he had never been able to trust his own ability to bring in enough money to support a family. He hid that psychological burden under a superman persona and frequent assertions that he could always supply the essentials. I had grown up with the blarney, and naively assumed that he meant what he said, that our emotional bond was as strong as I wanted it to be. It had taken me a long time to see fragility and vulnerability underneath my father's rippling muscles.

I have long viewed both of my parent's lives as being essentially tragic, and I don't blame either of them for anything. The efforts of so many families around the world turn out similarly, and often much worse. Nevertheless, I am trying to use my parental family experiences to understand what more generally goes wrong with human potential, and to deliver that understanding to the future.

Chapter 4
Brain Resonance

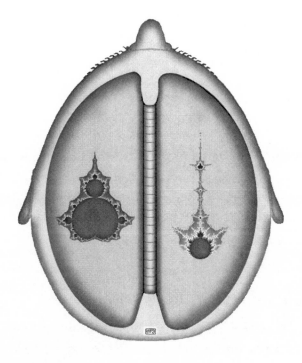

Our Hidden Skills

There was a man who came upon the word "prose" for the first time, and looked it up in the dictionary. He was astonished to discover that he had been speaking it all his life. In a similar vein, each of us is a neuroscientist. We are self-taught brain jockeys driving a body around in the fabric of reality. If you have read this far, you must also have long ago learned to read short sets of black squiggles and to comprehend what they correspond to in the real world—a stunning achievement!

To understand the present chapter, you will require another set of amazing skills. Here is a list.

1) The ability to form loose associations between prior experiences—you were doing this when you held dead rabbits, electric rabbits and greyhounds in mind while a story unfolded.

2) The ability to turn loose associations into verbal rationalizations. Fred did this when he decided, "I'll tell Alan that I threw Albert to the greyhounds; he'll love it!"

3) To learn from repeated failures—like all those beatings.

4) To revise your attitudes when the context changes. No matter how much you love lizards, there is no point capturing them when they are all over the place.

All this list does is make these skills a little more self-conscious, so you will be able to recognize what I'm on about a little later. Like right now.

Korbinian Brodmann: The Patchwork

You develop and verbalize loose associations by manipulating electrical patterns within your brain tissue. It's worth knowing the story of how we ever developed that perspective.

In 1905, a German neurologist, Korbinian Brodmann, peered into a microscope and examined the surface of a mammalian cerebral cortex. He noticed that discrete patches had a relatively constant pattern of neuronal relationships, but, across discernible borders, new patterns began in different patches. When he drew what he saw on paper, he ended up with dozens of zones in a map. It was reminiscent of the personality zones of old-time phrenology, which had discredited itself by prematurely classifying some bony skull bumps as predictors of criminality.

The good Korbinian was not about to claim that he knew what the patches were 'for', or what they predicted; he simply numbered them. To give him credit, we now call them "Brodmann numbers." (See Garey, 2006.) We had already learned that nerves work by carrying electrical signals, but we now needed to know more about the particular electrical patterns associated with each Brodmann patch. But how does one study electrical behavior, the neuronal language, of constituent cells that are too small to see with the naked eye—and which even overlap in the field of view of a microscope?

When the first breakthrough came, it was surprisingly prosaic. A glass capillary tube can be flame-softened under tension. As the glass within the flame melts, the hot region can be stretched. Surprisingly, the central hole survives, now defined by a very thin glass wall. After breaking at the thinnest point and filling the tubes with salt solution (which conducts electricity), one can place the tip near and even through a neuron's membrane. Even though one can't quite see what one is doing, the cell membrane, which is a flexible double layer of lipid molecules, does the trickiest work for you: even as the glass penetrates the cell, the lipid seals the interface. Consequently, the electrically conductive salt solution can report the voltage from inside the cell all the way out into the laboratory.

Although such a tiny lance would be a miserable jousting tool, one of the most famous of these microelectrode-bearing physiologists, Bernard Katz, received a knighthood, along with his Nobel Prize, and so he became Sir Bernard of the Microelectrode.

If the output wires from the glass parts are connected to an oscilloscope screen, the electrical fluctuations within the neuron become visible. If connected to a loudspeaker, they become audible. The pioneering studies were done (by Sir Katz) recording from muscle fibers at frog neuromuscular junctions.

When the studies were extended to the cell bodies of neurons in brains, many different firing patterns were discovered. It seemed worth knowing which neuron in the brain produced what pattern? Another inspired soul realized that, at the end of each recording session, a burst of lethal electricity could be used to force dye out of the glass pipette. By subsequently slicing the brain into histology sections, mounted on glass slides, the single bright yellow neuron, the source of the signals, could be traced. And so a database of which neurons fire in what ways was built up.

Over the decades, microelectrode studies became increasingly sophisticated. Clusters of glass electrodes were made so as to record from many neurons simultaneously; they were lowered down onto Brodmann patches. One could even lower a second electrode-cluster onto another patch, recording both sets simultaneously. In his books *How Brains Think,* and *The Cerebral Code* (2006) William Calvin, describes how a pattern arising in one Brodmann patch attempts to invade its neighbors. The attempts are usually unsuccessful, and the second patch goes back to doing its own thing. The next point is worth emphasis. The most interesting things that happen in the brain, the magnificent capabilities it displays, are not necessarily the result of its usual behavior. As I've just mentioned, pattern invasions usually fail, and presumably have no further consequence. Occasionally, however, the traveling patterns find reinforcement in the new patch, i.e., resonance occurs. The search for resonance may be the basic form of 'trial and error' that I am emphasizing.

Calvin suggests that the patterns that dance across research worker's oscilloscope screens have meaning within the brain; that their individual searches for resonant interactions are equivalent to searches for loose associations. Once an association forms, it has become a combined pattern, and it can start another pilgrimage around the brain, seeking additional reinforcement.

How many times this recursive repetition can happen is anybody's guess. A few pages back I noted an example: whenever I hear Mick Jagger sing *Factory Girl,* the sounds reactivate memories of a particular girl in a particular factory.

An Evolutionary Fairy Tale

I now want to use a fictional story suggested by Calvin's *The Throwing Madonna.* It will help you recognize how new thoughts may have erupted in the past.

The upright apes of days gone by had long arms and legs. Relative to their ground-scurrying cousins, their upright posture had relieved the shoulder socket of antigravity duties, and this allowed the joint to reduce in mass, which made it easier to throw things.

When predators had come prowling, the tall apes had instinctively screamed while flinging dung, sticks, and then stones. Forceful, accurate stone throwing drove predators away with pleasing efficiency. When the campfire burned low, however, wolves and big cats came prowling—and they could drag off a youngster.

One evening, a wiry haired ape, Albert, collected an arsenal of stones ready at hand, before bedding down. Thoughts favored in one context had successfully discovered a second home. The other apes did not see the method in the madness until the big cat came, and Albert drove it away alone. After which he got laid. The younger males examined the sequence of events carefully in their own minds. Some began to collect arsenals of stones before bedding down at night, having evidently experienced a recursive insight.

One day, their forest home was hit by lightning, and caught fire. They survived by retreating into the river. During the following night, Albert awakened with hunger pangs gnawing at his belly. He took a walk in the moonlight. By now, he never went anywhere without his favorite defensive rock in hand.

A rabbit was grazing gratefully in a patch of surviving grass and it looked up as Albert approached. When he stopped, the rabbit bent back to the short foliage, but Albert knew it would run if he ventured closer. As he watched the rabbit, however, remembrance of rock throwing wandered from a patch of brain tissue near his left ear, and found itself settling into another patch of neurons already stirring in response to hunger, a patch over his left eye. The pattern of the old resolved need for defense resonated with the current desire and ignited variations on the theme. Ideas were breeding. A new and strongly amplified pattern shot across from one hemisphere to the other. It told him of another context in which rocks might be useful.

Albert perceived the effect as a sudden epiphany—a new view of how the world was arranged. The solution to his problem seemed so simple now. Self-propelled, four-legged fruit might be plucked, after all. He felt the weight of the rock, chose the moment and pitched. The rabbit's head rose to meet the impact, and it was still thrashing when Albert knelt and took it in hand. Those front teeth looked unpleasant but its whole head came off with a twist and a pull; it was just like plucking fruit from a tree. Life presents strange pattern repetitions.

He gratefully drank the rich, warm juice pumping from between the shoulders. Then he shifted his harvest to his left hand, gently bent and retrieved the good stone. He cradled the teardrop shape in his best hand, at the end of his best arm. His teeth gleamed in the moonlight. His companions were again amazed. He got laid. Albert got laid a lot. (This was an adult fairy story.)

David Hume on Thinking

The English philosopher David Hume declared in 1748 that we cannot comprehend anything except with reference to previous experience. Only after you have first experienced yellow, can you think of yellow objects.

The significance of this venerable truism will permeate the rest of the text, so I want to emphasize it a bit more. It means that a phase of experiencing must always precede thinking. For example, I experienced the force of sexual attraction at the top of a ladder in Fort Myers. Afterwards, seeing a pretty girl, I would think about.

The converse is sometimes easier to see: in circumstances that have no resonance with the past, you will be hard pressed to think any thoughts. Instead, you will find yourself in a highly charged experiential mode. By the time you become an adult, however, such first time experiences will have become rare; instead, nearly all experiences will overlap with something in memory.

That was the good news. Now consider this: by adulthood, nearly every experience will actually resonate with many earlier experiences. Probably too many. Consequently, the problem for the brain will shift: Instead of searching for 'something' in the past that is usefully relevant, the challenge will be to discard the marginally relevant and mostly useless resonances so that more appropriate guides to behavior will stand out.

Musical Resonance

Steven Mithen's archaeological epic, *The Singing Neanderthal* (2005), suggests that music preceded language; that singsong and crooning to soothe crying babies reflects that ancient heritage. I am adding the notion that musical appreciation may be rooted all the way down in the ability of neurons to entrain each other in resonant harmony and to recognize dissonant conflict.

Individual enthusiasms for different musical styles often differ; clearly there is a gatekeeper function that is style-sensitive and that my speculation does not address. If neuronal assemblies have a fundamental skill of resonance detection, however, then music is neither an epiphenomenon nor icing on a cake.

Musical appreciation may disclose the most fundamental behavior of neuronal assemblies. This consideration may explain the potential of music to feel deeply meaningful: it supposes that the search for the tonic is homologous with the search for meaning, i.e., that music exercises and rehearses the brain's resonance technology for extracting meaning. (Ref: Howard Gardener cites anthropologist Levi Strauss as an original thinker of such thoughts.)

Poetic Resonance: Rhyming

The poetic technique of rhyming provides another literal example of the brain's resonance sensitivity. The decaying reverberations of the sound of one word amplify one's sensitivity to a similar sounding word arriving in a later line, sometimes several lines distant.

Stereo Vision

When we alternately open and close each eye, the midline view of the outside world shifts from side to side. Yet with both eyes open, we seem to see a single midline view. To generate this impression, the brain must present a synthesis of the overlapping elements to consciousness, and it does this very well. Stereo vision is therefore another general demonstration of the brain's skill at handling paired patterns. While children are getting the hang of depth perception, which takes place well after birth (Almli, C.R., and Stanley F., 1987), they are also developing a filing system for the episodic memories of the personal narrative. As we have noted, these will be used whenever we mentally compare something from the past to the present.

In recent years it has become evident that, when we recall old experiences, they reactivate the parts of the brain that were involved in the original perceptions. That reactivation suggests that the brain may be treating past and present as a stereo-pair across time, "seeing" if the past holds meaningful relevance (by which I mean resonance) with the present, and acting accordingly.

Although I have made no attempt to unravel reductionist detail in these overview considerations, I experience them as palpable leaps in understanding the restless, roving electo-rhythms of the brain. Moreover, with such broad thoughts already in mind, it is easier to absorb relevant reductionist details as they become available. Sigmund Freud was a master of the art of developing broad *a priori* speculations about mental activity, as illustrated below.

The Interpretation of Dreams

In his book, *The Interpretation of Dreams* (1913), Freud presents a dream analysis illustrating resonances of remembering. A female patient dreamt that she was wearing a unique hat as she passed a troop of soldiers. The brim hung down much lower on one side. She remembered feeling that she was protected from the soldier's carnal ambitions by being a married woman.

Freud has been roundly criticized for exaggerating sexual allusions—sometimes a cigar is just a cigar. In this case, when Freud asked the patient for any associations she might have with the hat, she offered none. Perhaps it was just a hat.

Freud suggested, however, that the central peak might be a phallic symbol, while the drooping sidepieces might represent testicles, which would indeed symbolize that she lived "under" her husband's sexual priority. The lady was taken aback and specifically sought to withdraw the claim that the brim was asymmetric.

Freud, presumably gentle but persistent, suggested that such details were important to dream interpretation. He felt she must have a loose association in her waking life that might offer a reasonable correlation; that with a little thought, she might discover it.

After a few moments of silence, the lady said that one of her husband's testicles hung much lower than the other: she asked if this were so in all men?

The point: if brain tissue operates by pattern analysis and seeks pattern resonances, then it will inevitably and automatically detect hat profiles that resonate with one's husband's undercarriage.

The User Illusion: Cutting Consciousness Down to Size

Tor Nørretranders, a Danish science writer, wrote a book with the above title in 1991; translated into English into 1998. He emphasizes a piece of textbook knowledge from Manfred Zimmermann at Heidelberg University: The information that flows through consciousness is about forty bits per second—about a million fold less than the total sensory information flowing into the brain.

In our adulation of human consciousness we typically ignore the fact that it becomes filled with the effort to remember a telephone number long enough to dial it. It compares to the slit in a concrete pill box, before which there is a ticker-tape summary of one's neuronal stock market, but with only one stock price being quoted per second.

On the other hand, given the extent of evolutionary trial and error behind the evolution of consciousness, forty bits per second has apparently been found not just adequate, but more probably optimal for survival. After all, there are only so many things we might do per second and perhaps doing them effectively requires freedom from information overload.

The widely noted invisible gorilla on a basketball court further illustrates a normal capacity to keep the contents of consciousness matched to that which we wish to perceive (*The Invisible Gorilla,* 2011 Chabris and Simons). A huge social price is paid, however, when we do not wish to perceive the misery incurred by genocide, torture, slavery or miserable working conditions, for we can easily keep our narrow window of consciousness completely filled with self-centered priorities.

Chapter 5
Round Trip to San Francisco

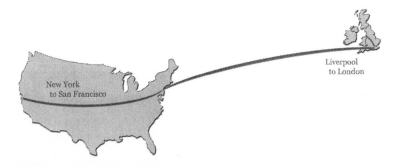

The Sniggery in the Mind

Each lecture I heard at Liverpool University was presented within the narrow frame of reference considered appropriate to its material. For example, a biochemistry lecture might have described the individual components of a metabolic cycle of food molecules. Although I had no spontaneous fascination for metabolism, I would rewrite my notes in the evenings and make more careful diagrams. Mother supplied coffee and cake.

For the bulk of the academic year, each lecture remained isolated in a separate mental compartment. Some were forgotten. It was the looming occasion of the final exams that forced a springtime review process. This became more and more intensive until I experienced the same thrill that I had first associated with 'knowing' how the electric bell worked. In these springtime marathons, however, an academic year's worth of lectures were being converted into a forest level matrix of comprehension—at which point some mental Rubicon was crossed and the burden upon memory suddenly lifted.

In other words, I no longer needed to 'try' to remember what was written on a particular page of my notes; I could 'allow' broad comprehension to pick out relevant details. Those details could still be thought of as regurgitations from the lectures, but the comprehension that elicited them now had a more profound quality.

I didn't know then that I was activating the different skills of the two halves of my brain. Today, I can be certain that I was. Nevertheless, deducing which hemisphere carried the baton during which phase of study—and labeling the handover moments with any confidence—is still hazardous. For the moment, I will only offer a subjective summary: the repetitive evening process (motivated by the desire to get a better job) eventually generated an internal Sniggery of cross-correlated patterns. I could traverse its paths with pleasure and even forge new ones whenever that seemed desirable. My thinking was no longer limited to simplistic verbal one-liners.

An Achilles' Heel

The choice of biochemist-to-be involved some mandatory courses. Zoology was a delight; dissecting various beasties was fascinating; learning in mixed company was welcome. The chemical consequences of the way electron clouds determine molecular assemblies were interesting and I was impressed that human beings had managed to unravel them.

For the third major, there was an option of physics or botany. I chose physics. This was a near catastrophe because my calculus wasn't up to the equations for bending of beams, which must have played such a rich part in the success of the Royal Fleet against the Spanish Armada. I'm being sarcastic: a math deficit is genuinely unfortunate. Since there were others in a similar dilemma, we were offered remedial classes. They didn't help. Some fundamental question in the foundation of calculus had not been dissected for me to see.

As a result, I couldn't even formulate the questions that would enable the teacher to see where my problem lay. I now suspect that it was a simple matter of inexact initial context understanding, in the absence of which repetition of details did not lead to insight. Nevertheless, I had to pass the physics course if I wanted to continue at university.

I chose a fallback position: I tried to understand the verbal descriptions of physics as deeply as I could, memorizing the final equations enough to write them down, but not embarrassing myself with inadequate attempts at exercising their powers. This got me the necessary passing grade, which allowed me to drop the subject like a hot potato, but not without some real appreciation for its content that I still draw upon today.

Fred and the Warming Trays

While I was still immersed in physics, Fred found another technological bandwagon to which he hitched his hopes. Breakfast "warming" trays had been invented. These were aluminum resistance strips bonded to glass plates. When plugged into an electrical outlet, the strips imparted gentle warmth to the tray, and wall panel sizes could be used as room heaters.

Fred conceived of their use in Merseyside hospitals where patient nightgowns were rather frequently set on fire by radiant heaters equipped with inadequately screened elements. The aluminum strips didn't reach the ignition temperatures of paper, nightdresses, or dust-bunnies. This was a good solid idea.

With the support of a drinking buddy, a sales representative for a distillery, Fred made the rounds of the local hospitals, talking to purchasing officers about possible interest in this technology. They came home well pleased with the impression they had made.

I don't think either of them understood the nuances of an international business, and I wasn't going to business school, so I didn't either. Such considerations became irrelevant, however, when they reviewed their sales pitch with the family physicist.

The strongest selling point seemed to be that the glass units didn't burn oxygen, "unlike those electric heaters where you can see the coils burning." In making this claim, Dad's rationalization mechanism had equated the glow of combustion and the glow of a resistance wire carrying electricity. From there he had formulated a creative but blarney notion that oxygen was better left for the sick patients. Although this had struck a chord with purchasing officers, it could not satisfy a test against broader knowledge. Resistance wire gets hot from the passage of electrons against resistance, and when metal gets hot enough, it glows. The glow will happen without oxygen; an example is the common light bulb 'burning' in a vacuum. In seeking to further their understandable ambitions, they had generated and verbalized an invalid loose association. I explained the reality, and earnestly encouraged reliance on the valid fire safety argument.

But the specter of Peat Bog Mary was stalking grimly in the recesses of Fred's mind. She seemed to be looking out at me, mortified again, and her son's ego was likewise crushed. I still remember the sensation of wanting to turn the clock back, to have them talk to me before they set out; to tilt the pinball machine; to alter the course of repetitive destiny. It was too late; the Gypsy had lied.

America Redux

In the spring of 1963, a notice appeared on a bulletin board in the student's union. It invited students to fly to America for the summer vacation. Living at home, studying every night and regarding poverty as normal had allowed great frugality; I could cover the cost.

I decided to make the trip. By this time, Fred had become a naturalized American, and I arranged a green card through his sponsorship. I was excited beyond measure. No more factory work for me! I borrowed money from the local bank, to tide me over in case it took some time to find a job. The English currency converted into $100, and Fred bought me a ninety-day pass for Greyhound Bus travel. The unlimited nature of the pass meant that one could sleep all night on a bus going somewhere distant instead of paying for hotel accommodations. Although this seems desperate in retrospect, I found it quite reassuring at the time.

Overland from Harlem

The last thing Belle said to me as I left the house was, "You won't go and see Frank, will you?" Mother's memories of Uncle Frank were not fond ones; she constantly referred to him as the black sheep of the family. As far as I could tell, this was because he had not married, had lots of girlfriends, and always had money to spend. Oh yes, and it was he who had persuaded Fred to work in America.

Frank was now the West Coast representative of the Seafarers International Union and was based in San Francisco. A second 3,000 mile journey from New York to San Francisco seemed too much to reach for, so I assured Belle that I didn't intend to travel that far.

Since a first night of lodging in a New York hotel was included as part of the travel package, my seat mate and I decided to spend the evening determining if the butchers, the bakers, and the candlestick makers of Harlem were indeed black. They were.

Lo and behold, the white boys felt uncomfortable with the tables thus turned, and we retraced our steps feeling slightly chagrined. On the way back from Harlem, feeling hungry, I bought a hamburger and coke from a street vendor. These provisions cost eighty-five cents.

This meant that a hundred bucks was a smaller fortune than I thought. Consequently, I wrote a postcard to Frank the next morning, informing him that his nephew, now twenty, was on his way overland by Greyhound, and would be looking for a job.

While an undoubted improvement over a wagon train, the cross-country bus marathon was grueling. I learned to stay away from the hot engine in the back seat, despite the convenience of being able to stretch out there. Food seemed so expensive I chose to starve. The scenery of America made up for that; it was continually amazing. I was particularly flabbergasted at the spectacle of Salt Lake City as we rounded a curve and saw it sparkling on the plain.

We stopped, and I noticed signs calling for sheepherders. When I enquired, I was told that those jobs were just for Mexicans; I supposed it was some sort of affirmative action program and stayed on the bus. After three days, and in the middle of a Friday afternoon, I arrived at the Seventh Avenue Greyhound Station, in San Francisco. A short walk to the Union Hall taught me about panhandlers. Fortunately, I had arrived just before Frank left for the weekend, and he welcomed me very graciously.

We set off for his apartment in San Rafael, where he had a Boston Whaler powerboat and his own dock slip. Within half an hour, we were out on San Francisco Bay catching striped bass. Well, Frank was catching bass. I was heaving up on a three-day empty stomach. Fresh broiled bass and a night's sleep renewed my feeling that this would be a great summer.

San Francisco Bus Boy

Frank wasted no time in telling me that my beard and sandals would make it hard for him to get me a job. I shaved, borrowed a pair of shoes, and was soon a busboy in a Geary Street restaurant.

The maitre d' was named Louis. He was a swarthy mustachioed character with a fine head of black hair, a scar down his cheek, and an apparent glass eye. I've been forever amazed at how effectively the context of a formal tuxedo made him seem sophisticated.

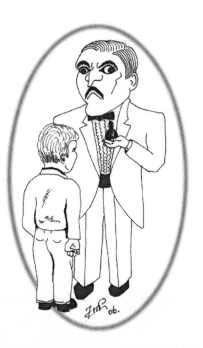

I worked there for four days, growing increasingly puzzled at the repartee, which, to my parochial worldview was incomprehensible. Furthermore, he had a penchant for giving me perfume sachets, so I would "smell nice for the coostomers."

Nor could I conceive of the reason for the frequent, redundant trips in which a waiter would ask me to bring an item up from the dark cellar. Said waiter would then follow, for further unclear reasons. The general atmosphere reminded me of something I'd felt before, around the Penguin people.

My frustration peaked when two men at a table were referred to as "Lady Godiva and Queenie." I had protested that they were guys. Applying a double gobsmack to his cheeks, my waiter said, with strangely exaggerated intonations. "Ooh, you are green, aren't you?" Coming home on the fourth evening, I put the puzzle to Frank, and learned that there were further depths to my sexual ignorance.

Benny Hill dressing in drag is not simply a leftover comedy routine from Vaudeville days. It is a reference to a sexual orientation in which like prefers like—and sometimes dresses as the opposite. I was working in the heart of a gay community.

My reaction? I had the typical defensive male impulse to bash the offending party on the nose. (Only now do I see this as difficulty with context revision.) Frank patiently explained that San Francisco had worked all this out already, concluding that one was entitled to turn down unwelcome overtures, just as heterosexuals do, but violent responses were frowned upon. We decided that I would be better off as a short order cook, a more heterosexual division of the service industry. And that's how I ended up in the East-Bay Transit Station coffee bar, slinging hamburgers while learning to drive Frank's latest Rambler.

I couldn't help but notice that life in the fast-food business, even forty years ago, improved upon factory work in Liverpool only in its avoidance of stoop labor. The lifestyle of these workers, living in cheap hotels, and taking vacations to see their families at the end of a long Greyhound Bus ride, was rather less desirable than the position of their counterparts in an old, small country with an extensive railway network. One can make a strong argument that centuries of reworking the infrastructure of England have made a mundane life more humane. For a further example, consider the "lounges" in English pubs. Homes in city neighborhoods were often too small for a lounge, so the local pub functioned as a communal equivalent. If you held short of alcoholism, the sociability of a cozy pub in every third block was a major social benefit.

The staff at the snack bar turned over with some regularity. Whenever anyone left, there was intense activity of the kind that David Brooks in his book, *The Social Animal* (2011), calls 'status sonar.' It focused on the quality of the new position being accepted. Was the departee climbing or falling further?

I stayed for three months, quite content to get up at four-thirty in San Rafael and catch a bus to open the coffee bar at six, and to wander Market Street on my way home in the afternoon.

When I left, I still hadn't revealed my student status, afraid that definite impermanence might affect my employment. So when news circulated that I had quit and would be leaving that day, the chief cook drew me aside for a confidential declaration: "You'll be back."

Ordinary Seaman

Frank invited me to return the following summer. By then he had received my seaman's papers, which meant that he could get me a job on a ship, as ordinary seaman, wiper, or mess boy. This was 1964, however, and civil rights issues were in the air. It was unclear whether Congress would enforce the right of blacks to work at jobs that had been previously reserved for whites. Frank's union was handling the uncertainty by dragging their feet; the rationalization that they offered was that they were full up, needing no more workers of any color.

Frank pulled strings, and I sailed on a Matson passenger ship, Lurline, whose labor was generally provided by a different union. The fact that white youngsters got union jobs broadcast an embarrassing system of affirmative action for the white in-group. All I can do is admit it. There were several of us in the same boat, as it were, and this provided someone to talk to, as we sailed from San Francisco to Hawaii to Los Angeles and back—an open triangle. I was the junior ordinary seaman and it fell to me to wake the next watch.

I was given more instruction on waking up seamen than on any other facet of the job. The cardinal rule is a negative one: never touch the sleeper. There was no need to theorize about psychological tension building up when confined in a small space with other testosterone-poisoned, territorial apes—it was quite palpable.

There were four in our fo'csle, in a pair of upper and lower bunks, and a brass plaque on the door allowed as to how this could be increased considerably in wartime.

In these confined spaces, the only territory one could 'own' was the few square feet of a bunk. When a sailor was in there, you didn't touch him or his bunk. When he wasn't there, you still didn't touch the bunk. Disregard for this warning resulted in physical harm for which no recourse was offered; one talked the next watch awake or suffered the consequences. Usually, that wasn't hard; seamen at sea sleep light, as though they are paranoid, which may also explain the violence with which an arm resting on a blanket can suddenly lash out when touched.

We sailed for five days to reach Honolulu.

I was on the four-to-eight watch, working with an able seaman from my foc'sle, Con. As dawn broke, we removed the net from the swimming pool and hosed off all the soot that had come from the funnels during the night. Then we filled the pool with water pumped in from the Pacific.

Second watch, afternoon duties, involved sweeping cigarette butts from the pool perimeter, while being forbidden to chat up the bikinied lovelies. There was little other work to do on a well-kept ship, and we either played cards in the sailor's square, or hung out in the bow, which was off-limits to passengers. We had movies in the evenings and almost nothing to spend our wages on. Gambling, however, was allowed in international waters. This opportunity left some sailors unable to leave the ship; they made trip after trip trying to pay off losses at the tables.

Reaching Hawaii, we saw native divers traditionally diving for coins in the clear water among hammerhead sharks, traditionally swimming with them. Once tied up, an exertion that required most of the crew, we had four hours of shore time, so I took a taxi to the beach, hired a surfboard, and tried to learn to surf.

This proved beyond me. I'm still pissed off about that. I have a cousin in Sydney who won medals for surfing, so my incompetence is unlikely to be genetic. I think it's because I can't swim worth squat either, and blame the Mersey River. That's where I learned, off the sands at the Seaforth Beach in the days when there was no wussy talk about pollution. Rivers were obviously for swimming in, so we did. In those postwar years, it was also where the sewerage ended up, and fairly quickly at that. Consequently, I learned to swim with my head held high above the water. It's hard to travel laterally if you're squeamish about your fellow travelers. Why we didn't die, I still don't know. I understand that a similar puzzle exists about the River Ganges. (Salmon have now returned to the Mersey!)

Lifeboat Drill

The trip back to the mainland was enlivened by a pickup at sea. A sailor on a nearby merchant ship had appendicitis and needed immediate surgery, which Lurline was equipped to provide.

The two ships came together in mid-ocean, crews lowered a lifeboat and they made the transfer with impressive competence. Fortunately for all concerned, I was just watching.

Later, in San Francisco, we had a lifeboat drill to develop such skills. Con climbed into the bow of a lifeboat at the upper deck level; I manned the stern. As winches turned on the deck, we were slowly lowered towards the water on a pair of stout cables.

I noticed two pieces of funky bronze engineering in the spine of the boat. Growing bored, I stretched a hand towards the one that was nearest, while asking Con what it was for. He whipped bodily around and yelled, "Don't touch it!" "Sorree," thought I, withdrawing.

A lifeboat's transition onto the water must be made at the same time for both ends. To do that, the lifting cables have to be capable of easy release. The attractive piece of funk was my end of the quick-release system, but we were still high in the air—it was too soon to get the bottom wet. This is called "on-the-job-training." It is a problem that seems to turn up with some regularity in modern civilization. And that isn't the end of this little parable.

The next time we went out on lifeboat drill, again in San Francisco, Con and I took a boatload of crew down to the water. After releasing the cables, we attempted to row the damn thing in a circle before returning to the side of Lurline. The Golden Gate Bridge seemed to behave as a strange attractor, and I began to wish we had brought lunch. Instructions from the mother ship, yelled into an increasingly inadequate bullhorn, eventually brought us back to Lurline and her lifting cables.

On a subsequent drill, we were in a different style of lifeboat. This one was fitted with 'idiot sticks' instead of oars. The stick is a vertical bar fixed into each rowing station. By pushing it forward and pulling it back, muscular effort transmits to a longitudinal axle that passes out into the sea, where it turns a propellor.

As long as the individual doing the steering is at least half witted, the boat goes where he chooses. Since the rudder position denies room for two half-wits, who might conflict, the design is an astonishing improvement over the multiple oarsmen, floating version of democracy. In this lifeboat, we again had a crew, and with stick drive it was easy to do the requisite circle and get back aboard.

I was now on comfortable speaking terms with Con and had learned that he had been at sea during WWII. As we stowed the gear away, I wondered aloud what the hell it must have been like in wartime, when a ship was torpedoed and the crew had to row for their lives, perhaps in rough seas. I learned that Con had been sunk twice. On the second occasion, the crew managed to quickly lower the lifeboats and get underway. When the surfacing bow of a submarine appeared above the waves, they even positioned the lifeboat between the deck gun and the setting sun.

By the time the hatches were open and the deck gun was useable, they had pulled away smooth, straight, and hard to see. It seems that there is a reservoir of potential capacity to get along together, when the alternatives are sufficiently dire.

All Hands on Deck!

One night, the usual etiquette was ignored. I felt a forceful shoulder shake, followed by the horrible words, "All hands on deck! There's a fire aboard!" I staggered topside, still in my underwear, to find that we were off the coast of California, in the dark.

Deck lights illuminated the grey coastal fog, and a searchlight was pointed down near the waterline where an open hatch was billowing black clouds. Lurline seemed to be examining herself. Since freezing would precede either burning or drowning, I went below and got dressed.

Meanwhile, another white-kid-ordinary-seaman had learned that he was also part of the emergency crew, and fate had provided him with the only available oxygen mask that delivered the actual gas. A smoke-reducing fire hose had been thrust into his hands, and he had been ushered into the corridor from which the dragon's breath was belching sable billows. When he found the fire source, it turned out to be a storage room full of mattresses and office supplies. Leaving the hose, still streaming, across the threshold, he backed out. Although the material in the room was quite flammable, it was not intrinsically dangerous, but the bulkhead at the rear of the room formed one wall of the ship's fuel tanks. If the heat had split it, or popped some rivets, we could expect a serious threat.

An SOS had gone out. Close to San Francisco, it had not been hard to find a nearby ship, and we could see its lights as it stood by, prepared to take on survivors. The passengers were summoned to the lifeboats. No drill this time. But the crisis was sufficiently slow rolling that many passengers went back down below, packed their suitcases, and lugged them on deck.

The officers then had to adjust those passengers' perspective: there is no room in lifeboats for suitcases. It doesn't really matter how much you paid for a ticket. Although I smile in retrospect, this particular context revision had provoked no laughter from the passengers.

Once the flames were damped, we tossed still smoldering mattresses into the sea, for an hour or so, as we sailed on to dock in San Francisco. An FBI man began interviews and we all learned of a pattern: the hero who reports the fire is often the arsonist. When the detailed questions begin, the worm turns. In this case, the arsonist had been another ordinary seaman, an orphan who had been reunited with his father in early adolescence but for whom the reunion had not been all that he had hoped for; the net psychological effects were still wreaking havoc with shipboard life. How often is a miserable childhood deleterious to one's own children and, through them, the whole of civilization?

Foreground versus Background

The Greyhound trip back to New York that October finished with yet another salutary lesson. It was my turn to bring Aladdin's treasures back from foreign lands, and as I hailed a taxi outside the Port Authority bus station, I was bogged down with multiple bags. Out of the corner of my eye, I realized that a gentleman on the sidewalk had looked me up and down. Then, taking his cigarette from his lips, he ground it out on the pavement with some deliberation. Not all that alarming, at least so far. As he pushed off from the grinding foot and began striding toward me, picking up speed, however, I perceived his intent and the logic behind it.

I dropped all my bags, turned and faced him squarely, adopting a gorilla pose that would have better suited Uncle Frank. The stranger spun on a dime, and quickly disappeared in another direction. I was amazed at the speed with which revised context understanding can inform an appropriate reaction.

I was also surprised that my abysmal street record didn't show. Psychologists have learned that rats raised in stressful circumstances by stressed mothers become hyper-vigilant and distrustful. Such responses have survival value in stressed environments; they had served me well.

Epilogue

Many of my experiences on Lurline generated thoughts that have resonated in my mind many times through the following years.

(1) The passengers who embarked on the ship had chosen to leave dry land and spend time floating upon the Pacific Ocean. Nobody taught them the significance of that context shift, and they hadn't thought it through for themselves. Crisis brought it home the hard way. The modern cruise industry still teaches this the hard way.

Did any of us realize that the decision to spread consumer cultures around the planet would dramatically change the ecosystem? So much so that whole species would be wiped out? Though many remain in denial, in 2013, the ship of global consumerism is finally sinking. All of us are faced with the question: What beliefs, habits, and possessions do we honestly have room for, as we try to survive?

(2) Is there existential similarity between steering a life with two cerebral hemispheres and steering a lifeboat? Evolution has assigned human steering to a self that concentrates on considerations passing through its consciousness; this leaves a complementary role of context analyzer, or map-reader for the other hemisphere. Whether the latter has its own form of consciousness is an open question.

Under the strain of global warming, will we all decide to get along as the going gets tougher? I can't see any chance of that happening under the old hierarchies of dominance or those hoary religious dogmas that have informed history and preempted rationality until now.

Could our situation be rescued by describing so rational a goal of human life that it would prove to be mutually agreeable, earning the consent of the people in democracies and authoritarian societies alike, and thereby promoting harmonious rowing?

I suspect that something like this was in Moses' mind when he took his rock chisel up to the top of Mt. Sinai for a forty-day stretch. As you know, he came back down with ten commandments, proposing that everyone should agree on that much. I didn't see anything wrong with the general idea until I read Leonard Shlain's book, *The Alphabet versus the Goddess* (1998).

Shlain notes that the Ten Commandments were the first use of alphabetic script in religious culture. But writing and reading scripture inevitably stimulated the left hemisphere, wherein testosterone works its magic to produce a dominating male ego. The original scribes were evidently male. Their output changed the symmetrical relationship between the sexes of pagan times, as follows. Female gods were now omitted; God hisself wasn't even born of woman. Furthermore, he never even married, let alone screwed around the way Zeus had done. To top it off, the whole category of Womanhood was allied with the devil and blamed for Man's expulsion from the Garden. Perhaps that explains why, some thousands of years later, we can look back on Moses' initiative but note that it was flawed and didn't work. We remain in a spiritual, material and even gender cacophony of confusion.

Could we do better than Moses; could we devise written guidance of such obvious efficacy that it wouldn't need to be forced down unwilling throats? I suspect that the left hemisphere's general answer to that question would be, "Hell no!"

...but we have another hemisphere.

Chapter 6
The Leaving of Liverpool

Honors Biochemistry

After my third year at Liverpool, I found myself with a bachelor of science degree and an asterisk. The latter was footnoted to mean I had done well enough to be eligible for another year, this time in the Honors program, specializing in biochemistry, doing some research, and potentially entering the Ph.D. program, all financed by the government. By now, Fred had noticed that putting a kid through college was something most parents were proud of, so I don't think he was quite as disturbed as previously—and I was working as a summer seaman, anyway.

That year was a total joy. I think there were twenty-four of us. We had an early morning lecture in a Quonset hut and then retired to the basement of the red brick building to do laboratory work. The lecturers were excellent; accumulating insight continued to intoxicate.

An opportunity to do bench research was new, and we each had a separate project. Most of these involved setting up finicky apparatus to detect infinitesimal changes. After you've measured a change, you have to do it again—to make sure that it's not random noise, and that you get the same measurement every time.

Unfortunately it was (random noise), and you don't (get the same measure). So then you really start to figure out how the apparatus is supposed to work and what it's doing instead; a large part of laboratory science is trouble-shooting the equipment. Furthermore, the possibility that an equipment malfunction is the reality at the bottom of a seemingly incredible discovery lurks in all scientists' minds.

My project involved alcohol dehydrogenase, a liver enzyme that metabolizes and so detoxifies alcohol. Much food is naturally toxic, which is why it is useful to have a liver. Before enzymes can entirely remove toxicity, however, some transient intermediates are actually worse than the initial chemical.

For example, alcohol dehydrogenase converts ordinary alcohol—ethanol—into acetaldehyde. Fortunately, acetaldehyde is quickly further transformed into harmless vinegar, so that it takes quite a lot of alcohol to experience acetaldehyde poisoning. The retina, however, is also rich in this enzyme. When a significant dose of methanol is swallowed, it is rapidly transformed in the eye into the very fast-acting embalming agent formaldehyde. Desperate alcoholics drinking methanol can pickle their own eyeballs from the inside, becoming permanently blind.

This kind of background lore stiffens one's resolve to seek insight. So it was that I found myself in a refrigerated room with a bunch of cow's eyeballs from the local abattoir, learning to invert them on my thumb. In that configuration, I could dissect off the retina and homogenize it, preparing a cocktail of the enzyme. This is the classic *Grind and Find* process of biochemistry. The scene might have inspired a Gary Larsen cartoon, and I guess it also fits the classic mad-scientist image. I see it as little different from the consumer trade in butchered cow-parts.

Most consumers partially burn said parts then carve off small chunks. After they are sprinkled with taste enhancers, they are speared with a fork and placed in the mouth to be chewed with evident relish. According to social convention, nothing is said about the death of the long-lashed, brown-eyed beast from which these morsels are derived—but I doubt that the gentle reader behaves so barbarously. (Who, me, defensive?)

I Leave Home

The final exams that year held heavy implications. I hoped to earn a First Class Pass because I intended to finally leave home and make the permanent emigration to America—without staying to get a Ph.D. in Liverpool. I could see that a First might serve as a lifeboat, perhaps making return to the English Ph.D. program a possibility if disaster should strike in the U.S.

Six weeks before the finals, disaster struck on the home front. Fred had made a visit, spent his cash, and returned to the States. His next ship had sailed to Bremen, in Germany, but it caught fire while docked there. After being paid off with airfare back to the States, he had stopped in Liverpool, still wearing work clothes. Of course, he went to see the boys at the Gladstone Hotel, came home drunk, and flew out the next day.

The following week, new carpet had been laid on our staircase. Belle had wanted carpet for quite some time, but we hadn't been able to afford it. Although her mother and several of her sisters had seen their way clear to earning a living, Belle was adamant that Fred was supposed to be the support mechanism she thought he was when she married him. She declined to take up the slack. We had recently been so broke that I had borrowed money from my friend Arthur, who was working on buses, to tide us over. So I asked where the money for the carpet had come from.

Mother said she had taken the money from Fred's pants when he came home drunk, knowing he wouldn't be able to tell what he had done with it. By then I knew how an American seaman earned the dollars that had made their way back to us, albeit in frustrating dribs and drabs. An impression of the inevitable future of their battle marriage hit me with nauseating force. Decades later, it came to pass exactly as I had feared. I now see foresight as an example of pattern analysis by my right hemisphere. Accurate pattern analysis.

With this news, I simply decided to leave home. Fred bore physical scars that Belle had inflicted when she was enraged. He displayed them with a peculiar pride—they demonstrated his cross in life. I was concerned that my decision might trigger a similar episode, and composed myself to resist forcefully, should it be necessary. For several nights, I made sure my bedroom door was locked from the inside.

A room to rent was available with my classmates, in a downtown student lodging, and I arranged to move in. The first of several surprises came Sunday morning, when I went up to the kitchen to make a cup of tea and swallow some wet Wheatabix. Two of the girlfriends, still in their nighties, were making breakfasts for their boyfriends. I sat down at the table, head in my hands.

I remembered Billy and Joey and realized that, by living at home and concentrating on studying, I had overlooked a rite of passage for nearly four years. I quickly completed the necessary context revision.

Another benefit of leaving home was dining with fellow biochemists. We were all sweating the same exam, all heads full of the same dizzying array of chemical structures, facts, and latest explanations for how things worked. I learned of some errors that I had built into my more private study efforts, and was thankful to also revise those.

Final Exams

The exam itself was a five-day series of three-hour essays. The time for mental distillation had all ticked away at the moment that the Monday session began. We were in a large exam hall separated on each side by an empty seat. The next desk was empty; the one after that assigned to a student doing a different exam, and the fact that I didn't even know my neighbors turned out to be a blessing.

At precisely 9:00 a.m. we were allowed to turn over the exam sheet, and read the questions. As I scanned them, heart pounding, I realized that I could respond intelligently to the necessary four of the six questions. The wave of relief that swept over me was accompanied by a pronounced urge to piss. Deciding to get that over with right away, I lifted my hand and asked to go to the Gents.

Facing the porcelain pottery, my mind was apparently in overdrive, revving through the issues I would commit to paper. But the flow of neuronal electricity must have jumped the tracks and resonated in the wrong Brodmann patch of cerebral cortex. Although the specific pattern still held the appropriate information, it opened the wrong set of sphincters. There was an immediate weight in the seat of my pants. Like a soldier in battle, I had lost my lunch. Yesterday's.

My entire future was now hung up in my underwear. Adopting a crab-legged posture, I walked backwards into a stall, shut the door and carefully removed my trousers. Attempting even further delicacy, I took off the inner cloth and dropped the whole shebang into the shorter porcelanic idol. I attempted to clamp down on my emotions with military force, and returned to the task at hand. Could I not have just asked to come back later? Yes, a year later.

The rest of the week was an anticlimax. I was aware of crossing yet another sort of Rubicon. My years of sympathetically and fairly uncritically absorbing past wisdom were at an end; I was ready to be an independent thinker. The potty problem had been humiliating; it may have tipped some psychological switch.

The Interview

I think this sense of earned independence was already swelling when several of us were called for interviews, which included an outside examiner from Wales. I was on the list, which implied that I might be in line for a First.

I was asked about enthusiasms. I interpreted this to mean that really good students had enthusiasms, and felt an obligation to have one. So I mentioned insect biochemistry. (I should have said bullshit!)

This led the external examiner to ask me what the Queen Bee substance was. I had no idea, and realized I'd dug a trap for myself. I explained that I hadn't done anything about the supposed enthusiasm, being too busy turning my notebooks, full of trees, into a more familiar forest.

After establishing that I seemed to know nothing that I hadn't been taught, the visiting bloke asked me if I thought biochemists might have some special insight into student drug abuse? This felt like the sort of setup for which that Charlie Brown falls every time.

I could have rabbited on for ages, heaping praise on the training that had given me such wonderful insight into the chemical nature of the body, from which I might easily imagine the potential harms that psychoactive chemicals might do—but I didn't believe it. So I said, "No." What the hell, I'd been saying that for twenty minutes.

I finished up my perverse claim to erudition by suspecting that drug abuse was a psychological problem that was unlikely to be affected by a biochemist's preaching. My father now had a fully qualified biochemist for a son, and it wasn't having any effect on his alcoholism. We did discuss my intention to strike out for America. I wondered what psychological effect that knowledge might have on the committee.

I remember the turmoil around the bulletin board as we scanned them for the final decisions. I don't know whether realism or independence put me over the top, but I was awarded the First that I had hoped for.

I wanted to head off to America immediately afterward, to get the milk and money flowing, and to buy a car with reclining front seats. The institutional bureaucracy, however, insisted that I stay for the formalities of the cap and gown.

This ceremony included the delivery of the rolly paper (the diploma) by the head of the department, whom we had not seen very often. He apparently interacted more with Ph.D. students, but he did address us at the final graduation ceremony.

And so it was that I was sitting at Professor Morton's feet when I learned what the grand plan had been all along.

Epilogue: The Lipstick Mines

Apparently there was some residual uncertainty as to exactly what it was we were going to do for a living. Professor Morton's speech ended with the following thought: "*The cosmetics industry is adding many new chemicals to their products, and many of these are applied directly to the skin. I believe that a biochemist can be sure of a lifetime's employment, for example, checking on the carcinogenicity of lipstick.*"

I was kind of sideswiped and gobsmacked again. I'd spent all those years with the books, all those nights cramming to escape from my father's employment trap, and lipstick chemistry was to be my holy grail?

I had been born to get my mother out of a munitions factory, and now I was apparently destined to analyze lipstick the least pleasurable way. Was this the best that civilization, at least England's version, could do? I was glad that I had planned to escape to America.

Professor Morton's concern, however, proved to be well placed. Just a few years later, professional biochemists would be driving taxis, and I would be thinking of returning to the ocean waves.

(I recently learned that some manufacturers continue to put, of all things, lead in lipstick.)

Chapter 7
Vietnam or the Engine Room?

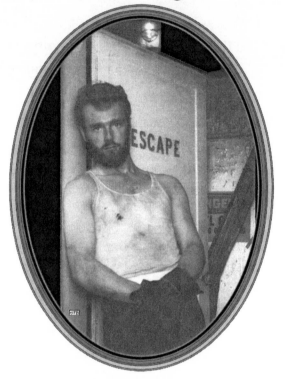

Vietnam and the Draft

After arriving in New York, I flew on to the one job interview that I had managed to prearrange; it was with Proctor & Gamble in Cincinnati. They were enthusiastically putting something called GL-70 into Pepsodent toothpaste; the presumed reason one would *Wonder where the yellow went.*

It was 1965. I had visions of a fat salary and serial carnality. The interview, however, revealed a more likely scenario, for President Lyndon Johnson had just announced his decision to escalate a U.S. effort to 'defend' South Vietnam.

This would involve a major increase in U.S. Forces, and so a single male college graduate was prime draft material. Moreover, immigrants were eligible for the draft six months after the date of their first arrival, and my immigrant green card was two years old. P & G told me that I should expect to be sent to war immediately, and that this made it hardly worthwhile for them to start training me. They were also certain that no matter how well I had performed in prior exams, failing to get a Ph.D. would limit my eventual options.

Although I was now a graduate, and presumably of value to somebody, my own capital was still in the $100 dollar range. At this level of frugality, it was important to save money, and I was able to rework the return ticket between Cincinnati and New York so that I could instead fly on to Frank's place, now in Long Beach, California. I had been told of the advantages of enlisting. As I flew westward, I pondered that decision. Since I knew that I wanted to stay in the U.S. and become a citizen, I provisionally decided to do so. The night I arrived, however, Frank and I watched a national news report, and it happened to describe the history of the Vietnam conflict.

In the late nineteenth century, France was second only to Britain in the proportion of the globe colonized, and the part of Asia known as French Indochina included Vietnam. But in the Second World War, Imperial Japan ousted the French and took over Vietnam. With the defeat of Japan, at the end of that war, France attempted to reassert control over its former colony. Under the leadership of Ho Chi Minh and with the support of the Soviet Union, the indigent population handed the French a massive defeat at Dien Bien Phu. As a result, an international Geneva Conference was convened, at which the French agreed to depart gracefully during a temporary two-year partition of the country into North and South Vietnam. That agreement included a provision that nation-wide elections would be held in 1956, to reunify the country under whoever won.

Watching the televised description of how America proposed to substitute for France in South Vietnam, and so turn an agreed temporary division into a permanent one, I sensed an extension of the rampant rationalization monster. I had a limited grasp of the horrors of England's colonial adventures, but I was aware that advanced-nation paternalism had been a disguise for exploitation.

Moreover, growing up in England after a war with Germany, I had developed strong critical feelings about the German people's patriotic support for all that Adolf Hitler had proposed. Now I doubted that Lyndon Johnson's policy could ever be made morally, or even practically "right," even by an apparent military victory. As I saw it, the body politic of America had ignored the ideals of its founders, and was now supportive of a President who, with a fresh round of rationalization, was applying lead-laced lipstick to a pig.

I wrote to my Honors program advisor, George Pitt, who was on sabbatical in Ann Arbor, Michigan, asking for advice. His quick reply explained the possibilities of getting a teaching assistantship to finance a Ph.D. experience in America. This would be coupled to a student deferment. He suggested that I get it done before the Selective Service System shanghaied me.

Singing the Wiper's Blues

I asked Frank to get me on another ship, pronto, so I could earn enough for a pilgrimage around potential graduate school programs. A Liberty ship was in the harbor, bringing steel from the East Coast through the Panama Canal; the captain intended to load lumber in Oregon for the trip back. She needed a wiper, and that was one of the jobs for which I was supposed to be adequate.

On freighters, viscous fuel oil and water are piped to fireboxes and boilers, respectively, and their steam output drives the engine.

Between them, these systems animate a genie—an enormous stainless steel worm. Revolving and screaming defiance, he passes through a stuffing box and out into the salty sea, where he is forced to turn a gigantic bronze propeller. This transition from inboard to outboard provides another opportunity for leaks. At any given time, one, some, or all of these systems will be dripping and there will be a need to wipe up the mess. This would be my job.

Frank came aboard and introduced me to the captain, who assigned me to a foc'sle and told me to get down to the engine room immediately. I suppose something was leaking. After changing into an all white ensemble of old trousers, tee shirt, socks, and sneakers, which would turn out to be a curious camouflage, I wondered how to get into the engine room.

The primary magic of ships is that they allow you to walk on the water without getting wet, but one can sail on a large ship, as I did on Lurline, and be unaware of the engine room mechanisms that make it possible to not just float but to go somewhere. Now I was about to discover the belly of the beast.

Oval doors were cut into inner bulkheads on each deck. Set about six inches above grade they were heavily locked with a circular wheel mechanism. I figured these precautions were to prevent seawater from getting into the ship's vitals when waves broke over the bows.

I arbitrarily chose one such door, revolved the wheel lock, swung open the oval plate, stepped inside and closed everything behind me. When I turned around, I couldn't see a thing. Furthermore, the heat and noise seemed incompatible with life. It was as though I had passed through the back of a wardrobe into another universe. Almost by reflex, I reversed everything I had just done and stepped back out, now slamming the door behind me. Appalled, I reconsidered... Vietnam or the engine room?

Despite my reaction, I knew that somebody must be in there, working, and somehow surviving. Moreover, the combination of moral and physical dangers In Vietnam still added up to the greater evil, so I revolved the wheel lock again, entered, and committed myself to explore the hidden realms of motive power. A narrow catwalk led deeper into the misty bowels and I reached out my hands to grasp its rails. *Too hot, forget that!* Stretching my arms straight out instead, I continued into the mist, my beard and hair curling up, while I hoped to see a staircase before I fell down it. I must have looked like supporting cast in a horror movie.

When the mist suddenly cleared, I could see several decks down, to a control board. Two men were standing there. The cleaner one was bearing arms with a cigar just pulled from his mouth; the grimy one held a Stilson wrench. Backs to the board, wide-eyed and slack jawed, they stared as I clambered down to their level. The cigar smoker, composure regained, said, "Who the hell are you?" I told him I was the new wiper. He gestured with his stogie, to where I had come from. I saw that a steam pipe below the catwalk had developed a small leak and was spraying a plume of hot, supersaturated mist just inside the entrance door.

"We heard the door clang twice," he said, "and then we saw two hands and a head floating out of the steam. We thought it was the second coming of Christ!"

Having established that I belonged on the other end of a non-celestial hierarchy, the Chief Engineer introduced Jesus of the engine room to a fresh mop and a bucket. Although I did squeeze some oil into the bucket, the mop became quickly saturated with black fuel oil, the main destination of which was inevitably the bilge. As Jacques Cousteau once noted, this is the stuff that, when the bilge is pumped, has left a black line upon all the cliffs at the edge of all the world's oceans. A massive graffiti.

The worst aspect of the job was called "blowing the tubes." The smokestack, a feature of steam ships, is an outer wrapping that protects a bundle of smaller pipes. These take combustion products from the fireboxes out to the open air. But the narrow pipes accumulate soot, which further narrows their bore and will eventually inhibit the intake of fresh air for the firebox. Consequently, a valve is fitted to each of the tubes, so that they can be steam-blasted clean twice a day.

The valves were opened and closed (on this ship) by a series of endless chains. Each of sixteen had to be completely rotated four times by the poor jerk at burner level. Said jerk was the wiper. During the first blast, the chain would be relatively cold but as the links passed through the valve assembly at the upper level, escaping steam heated them. By the second cycle, only a miracle could have prevented double blisters from developing on each finger. I proved incapable of exerting the necessary control over reality; even heavy gloves did not save me.

A Lesson in Money Management

My foc'sle mate was again of taciturn disposition; he spent most of his off-watch time quietly reading crude and rude porno paperbacks. I hoped that this would be my last time at sea and, since I had a Polaroid camera with a self-timer, I decided to document the misery. In the flash-lit gloom, the first prints were marred by pronounced redeye so I got some aluminum foil and took a wire coat hanger from the foc'sle locker. Bending the hanger into a diamond shaped frame, I crumpled the foil into a ball, before slowly unfolding it to fit over the wire.

My companion was in his upper bunk, reading the latest novel. Halfway through the unraveling, he said, "What the fuck are you doing?"

Treating this as an invitation to chat, I explained the problem with redeye and how the chaos of crumpled, reopened aluminum foil would diffuse the flash. I wanted to prolong communication, and so generated a question that might do so. Nodding towards the book now lying on his chest, I said, "Why don't you write one of those?" He said he had.

His early shipping experiences had inspired him to write a sea story called *The Decks Ran Red*. The script had been bought and used in a movie. If I remember correctly, Broderick Crawford had played the lead, and my foc'sle mate had received something like $20,000, which he had managed to spend in two weeks ashore—after which the Muse had deserted him. I saw an immediate lesson in that, a pattern I did not want to repeat, and it resonated much later, when my original hundred dollars had finally become something over a hundred thousand.

Next day we sailed for the Columbia River, where we dropped off the last of the East Coast steel, and then steamed up to Longview to take on tree trunks. Two weeks after boarding, and not wanting to pass back through the Panama Canal, it was time to unship. In unionized America, I was discharged with more than the equivalent of an English shilling. Consequently, I could afford to travel (by Greyhound) to every biochemistry department I could locate on the West Coast.

The exercise paid off in San Francisco's Medical School. The former chairman had died unexpectedly, and the department seemed fairly chaotic. Fall was approaching and they had more teaching assistantships than candidates to use them. Professor Fineberg, who interviewed me, was familiar with the English exam system, and felt that my record made me readily acceptable. I'm sure I also had to get references, but it seemed that my search was over. I took the trusty bus back to Frank's place in Long Beach.

Down the Rabbit Hole

There was time to goof around before the semester began, so one day Frank and I set out for Catalina Island. This day demanded more context revisions than any other I have ever known. It began when we left, in the dark of night, carrying extra fuel. The calm sea was suddenly lit by streaks of light in the water alongside us. Moments later, fish came flying over our heads!

Just before liftoff, as flying fish accelerate under water, they can disturb bioluminescent algae near the surface: these were producing the seeming tracer rounds in the dark sea. These fish also instinctively orient toward light, and our running lights led them to whiz over our heads, like so much artillery.

Frank wanted to collect some for marlin bait, and he had me hold aloft a big net, hoping that one would whomp into it, thereby catching itself. In the dark, however, it was hard to see them coming. When I did, I felt more like ducking than fishing, so we didn't get any. I wondered where the hell we would put a marlin, anyway. The boat was only just big enough for the two of us and all the extra gas cans.

When we reached the kelp beds off Catalina, in the early morning, we took to the water, one at a time, with wet suits, snorkels, and a spear gun. Should a shark take an interest I was advised, of course, not to let it know I was afraid. I discounted that wisdom and privately planned to retreat into the kelp. (This was an error—sharks aren't inhibited by kelp.)

As one hovers in the water over the extended green tips of a seaweed encircled island, there is a delicious impression of defying gravity; it feels like flying. At the boundary where the seaweed fronds rippled the surface, I tried an experimental dive. Some ten feet down, I noticed a tunnel. It extended toward the island.

I could see that a shaft of sunlight illuminated the far end. I should then have wondered how a tunnel maintained itself in a mass of kelp fronds—that would have been an appropriate acknowledgement of a mismatch between context and details. But I was busy feeling proud of myself because I was performing better under water than I do on the surface. As we are warned in the Bible, pride often precedes a fall.

Fixated on traversing the tunnel, I resurfaced to take a few deep breaths, dove again, found the passage, and swam into it. When I discovered that I was not alone, it was too late to turn around.

Something large in a white winding sheet with a pair of black eyeholes torpedoed towards me, for all the world like Caspar the Ghost. Suddenly feeling more like food, I turned my torso so that I backed against the weedy sides of the tunnel—the unknown creature did likewise—and we passed within touching distance. Breathing being the next imperative, I swam on, surfaced, and swung around to look across the kelp fringe.

Another head had surfaced out there. As we gazed at each other I was reminded of Tinker, my mongrel buddy from childhood, for it was the doglike head of a baby harbor seal. It had evidently made the tunnel through the kelp on its shoreward journey. We had both panicked, each probably afraid that the other was a shark.

Since then, I've often had a fantasy of stroking a seal's skin while it was still on its owner; I missed a perfect chance. Of course, the seal pup might not have had a reciprocal enthusiasm.

The next revelation came when Frank explained that big abalone lived on the rocks. He said they could be pried up with a screwdriver, so long as the first attempt was decisive: inherently tasty creatures evolve defensive tactics, and abalone respond to halfhearted prying by clamping down immovably on the rock.

As an ocean swell lifts the water over a rock, the attached kelp is drawn vertically; it then drapes back down as the swell recedes. At the right moment, one can make a short dive to reach and hold onto rock. As the returning water lifts the weeds again, one can go to work searching for abalone. Frank told me that if I managed to lever a mollusk free, I should apply it to the wet suit, which it would grip reflexively, allowing another to be collected on the same breath of air.

"Don't collect too many though," he said, "there's divers still down there, covered in abalone!" I contemplated that reversal of perspective, then collected my share of abalone (two) and returned safely. You can't do that anymore; too many people collected their fair share.

The Tasmanian Devil, a Bloody Big Fish, and Me

On the way home, in the afternoon, we saw a pair of dorsal fins, which we supposed to be pair of sharks out on a date. As we drew closer, it became apparent that each flick of the forward fin was reciprocated, with a delay, by the second fin. We then began to suspect that it was actually one enormous fish, and that it was asleep. Since we hadn't any bait, Frank started looking for another option and decided to use the spear gun (not recommended—I understand it is illegal).

He turned the boat controls over to me, and cocked his weapon. I came up on the rear fin slowly, from behind. I grew a little nervous. Although its body extended a long way forward, and we couldn't see a pointed bill, we could see that its back featured blue pajama striping. This meant that it was a striped marlin, a formidable foe. Frank said, "It's as big as the boat!" He sounded thrilled, and I heard a Barnum-and-Bailey prosody in his voice. I got a sudden flash of the headline he was imagining in *The Seafarer's Log*.

Agent Boyne lands Fish as Big as the Boat

Frank's daredevilry, while generating good stories in the long after, took on a different character in the immediate now. It was very obvious that this particular stunt might leave me floundering off L.A. with an angry marlin, and no particular place to go—especially swimming as I do. Nevertheless, I could see that Frank was excited beyond measure and restraint; he was back in the ring.

Fully aware that multiple hundred pounds of muscle would dive when speared, and that it would easily pull him overboard, Frank willed himself to pull the trigger and then throw a hitch onto a forward cleat, regardless of what happened next. (Stan Laurel would not have figured this out, which makes slapstick hilarious... from a seat in a theatre.)

When Frank pulled the trigger, and the bolt shot forward, my eyes registered its passage to one side of the fish. With a gentle twitch of its tail, the piscatorian Zorro sounded, leaving our dangly spear to be pulled back aboard. The champion wrestler was too busy getting the hitch on to notice.

When your garden-variety immigrant first witnesses the wonders of America, he gains an overwhelming impression that his every prior experience is being outclassed. It must have been getting me down a little. I'm afraid I used the occasion to soothe this feeling: "I can't believe it: how could you miss a shot like that?"

I was still giving him hell, when the fins resurfaced and resumed their lazy flicking. Der monster had not felt threatened enough to wake up! I choked off my invective with a silent curse, and we resumed battle stations. This time the bolt went deep into the fish's back, an inch to the side of the main dorsal fin.

I should mention here that the vessel we were in was Frank's third of the same model. It had a triple vee in the hull, making it roll resistant. While probably great for whaling—close to what we were doing—the overall span slammed hard against oncoming waves. Recognizing this, the seats had strong scissor suspensions to minimize spinal shock, and the manufacturer had been able to guarantee that the deck-to-hull joint was unbreakable.

Unfortunately, and regardless of sea conditions, Frank had only one speed: all out. Consequently, the scissor suspension was repetitively flattened, requiring our spines to do the rest of the job. In this way, Frank had managed the separation of the deck-to-hull joint of his first two boats—within the warranty period. For the third, the one we were in, they had revised the guarantee. Now it simply said that Frank Boyne don't get no more boats.

Let's see, where were we? Oh yes, I remember...

With a spear in its back, the marlin awoke. This time there was no lazy tail flick: it lifted its aft end way high in the air with an evident intent to go deep. The hitch having again been completed, Frank stepped back. The line began to uncoil and then accelerate away. In a couple of seconds, it was all paid out and then there was a God-awful, thunderous crash. In the still silence that followed, the hull to deck joint was seen to have remained intact.

Seconds ticked by, while our brains sought resonance with their memories. The blankness of my mind at that point is fascinating to me now.

I was in the midst of a natural experiment that would be hard to perform more deliberately, for the situation correlated with no prior experience. Consequently, no resonances stirred in memory, and I could formulate no notion of what to do next.

Frank appeared to be feeling the same way, for we just let time elapse until it was finally obvious that we should pull on the line. It came back easily; the prey was long gone.

Once we could see the spear, the recent events under the water became clear. A brass circle, like a curtain ring, is supposed to keep the barbs of a spear closed as it enters a fish. The friction of passage through skin should rub it off, allowing the barbs to spring open, which would have fastened us together quite effectively. But the ring hadn't been cleaned since the last time it was used, and so it was corroded wasn't it? Consequently, it didn't slide off, and the fish had developed enough momentum to forcefully reverse the spear's intrusion.

I abandoned the pleasures of the jocular critique and wondered what might go wrong next. I didn't have long to wonder, for something had finally begun to glimmer in Frank's memory banks. After looking over the side into the deep clear water, he abruptly sat down in the passenger seat, saying, "Those things can get really pissed off. I've heard of them ramming their swords right through the bottoms of some big boats. They sank, too!"

In the hemispheric shift that followed, the view from my mind's eye portrayed us, and the boat, from the side. Moreover, I could see under the water, too, where a big angry force of nature, having chosen the obvious weapon, was accelerating, pointy end very much to the fore.

Yet again, roles had turned on a dime. I pushed the throttle forward: we reached a plane, slamming waves and spines as if lives depended on it.

110

I've learned more about boats since then. Judging from the force we experienced as the spear slid out, I suspect that if its barbs had been open, then both the cleat and a chunk of plastic might have been ripped away at the bow end. If enough of the hull had been opened, we would have promptly flooded and sunk. Unknown to either of us, our two silly lives may have depended upon the frictional value of a film of rust. Trial and error. Mostly error.

Chapter 8
Graduate School in the 1960s

A Mood Shift in the Sixties

As is widely known, Joseph McCarthy became the poster boy for witch-hunts in the 1950s.

In retrospect, his self-righteous supporters were obviously suffering from fear-induced irrationality but they successfully blacklisted many college professors and Hollywood celebrities— even Charlie Chaplin! All this was accompanied by one-liner sneers similar to the older one about the only good Indian.

When it turned out that even the U.S. Army was vulnerable to the tactic, that bulwark of society hired a Boston lawyer, Joseph Welch, to confront the accuser at a congressional hearing. It was broadcast on live television.

This became complex psychological drama. During his preparation for the defense, Welch learned that one of his staff had briefly been a member of a legal organization that had defended 'known' communists. This individual forewarned Welch of this 'taint' and consequent vulnerability to one-liner logic. Welch decided to keep him out of the hearings.

On this basis, however, McCarthy dramatized the proceedings with the claim that communists had infiltrated Welch's office itself! As that phrase was discharged from McCarthy's mouth, Welch interrupted him, "Until this moment, Senator, I think I never really gauged your cruelty, or your recklessness."

When McCarthy tried to ignore the comment and continue his attack, Welch angrily interrupted again, "Let us not assassinate this lad further, senator. You have done enough. Have you no sense of decency?"

The live images allowed the nation to witness McCarthy's body language: his scowls, his evident slavery to ego and his consequently imperious derisiveness; they saw bombastic irrationality in its full context. From our point of view, I want to note that such a holistic picture is only perceived by the right hemisphere; this may explain why it had not come across when most observers knew McCarthy only through black and white scripts of news accounts read by the left hemisphere.

The hearings evaporated McCarthy's social and political standing overnight. The previously spineless Senate censured him and he began to drink himself to death. It was many years before Charlie Chaplin returned. Was all this really necessary?

As the decade turned, and John Kennedy was elected, a sense of a change for the better was in the air. Those of us who had accepted society's educational opportunities and had dutifully burned the midnight oil, now wanted to discuss everything about the future for which we had been haphazardly shaped.

Although open discussion was universally acknowledged as essential, those discussions were starting to reveal that American institutions, including colleges, the FBI, and the CIA, had violated America's declared principles. In the decades since, such suspicions have been amply confirmed. For example, the U.S. has repeatedly interfered in foreign elections, with consequences that undermined American ideals and interests into the present time. (Unfortunately, the context revision that is required to admit to such past bad behavior can be so aversive that some individuals deploy, instead, denial, even to this day. I am building the case that this is a half-witted mental strategy.)

In the iconoclastic atmosphere of 1963, the UC Berkeley administration experienced a delayed ripple from the fifties, in the form of a McCarthyesque whine about Communists distributing leaflets on campus.

Instead of replying to the critic that this was to be expected in a nation that practiced what it preached—free speech—the administrators responded with a move that McCarthy himself might have approved: they declared that political leafleting would be banned on campus.

It is probably accurate to state that the administration's actions were consistent with the opinions of the politically appointed Board of Regents. In other words, they were consistent with the authoritarian conservative viewpoint, in which the university is an extension of parental authority, and obedience is the only appropriate behavior.

But when Dr. Kerr, the Chancellor, responded defiantly to the student's objections, he was throwing down a gauntlet. Among the students, political enthusiasts of all persuasions understood that the American constitutional declaration that authority comes from the bottom up was under threat. Dr. Kerr was reprising the metaphorical role of the royal conservative—King George. But fresh-faced kids picked up the gauntlet.

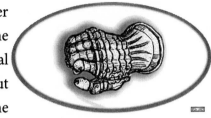

This was a straightforward First Amendment issue that one might have expected the press, especially the San Francisco press, to recognize the issues involved and so stand up for the student view.

The inevitable context within which the media compose their stories, however, is a meshwork of the opinions of advertisers and subscribers. Advertisers favor maintaining "order," and the subscribers were often working parents whose main concern was parochial: that their children get better jobs. In conformity to these views, the mainstream media implied that the students were behaving badly, and the preferred derogatory term was, "radical."

From the perspective of human history, American values are radical, but this irony is swamped, in many minds, by a deep belief that father-figure institutions should not be criticized on ethical or any other idealistic grounds.

In this fashion, Dr. Kerr was determined to ignore demonstrations. When the student body realized his rigidity, they responded with their own form of Tea Party: they marched en masse into the offices of the administration, and shut them down. For the administration, this proved the inherent badness of the students. They called in the police, and the students were arrested, en masse.

Note the fundamental assumption: civil disobedience against the authorities makes dissenters bad. Yet banning constitutionally approved freedoms to dissent is within the rights of those authorities. This is the asymmetry that America was founded *against*. Many citizens, proud of their country, substitute a narrow lapel pin for comprehension of that broad Jeffersonian point—a more cut-price form of patriotism.

Protesters and their faculty sympathizers, however, were able to insist that responsible governance of public higher education depended upon power sharing. This implied that faculty senates ought to debate university practice and to check administrative power when it was clearly excessive. The excessive threshold had been met by mass student arrests, and so the full Berkeley Academic Senate debated the underlying constitutional issue. When the vote was called, the Senate sided with freedom of speech. Political leafleting would not be regulated.

Those many parents who had been horrified to find that their children expected basic American principles to be implemented routinely were now left with a psychological difficulty: the workings of the system had declared them to be wrong.

It has taken me many decades to understand the extent to which the associated ego wound would fester and undermine the future. Those who found themselves on the losing side then were also on the losing side through the subsequent civil rights and women's rights struggles of the later sixties. Although they now rarely argue for reversal of those reforms, for example to get Gwen Ifil off Public Television, they still harbor as much resentment as Edmund Burke felt toward the French Revolution. Their rearguard actions include a long-running effort to eliminate public funding for Public Television. Of all this, however, I was still blissfully unaware when I came to San Francisco in the fall of 1965.

Glory Days in San Francisco

The student union at the medical center had a Rolodex file listing rooms to rent, and one was advertised for fifty-five dollars a month. It was beside Golden Gate Park, and within walking distance. When I dialed the number, I was treated to an extremely sexy voice and was invited to come right over to see it. Off I trotted, Hollywood images of the perfect landlady ricocheting through the bedrooms of my mind.

Ringing the doorbell, I heard heavy feet coming down a staircase and the door creaked open to reveal a very tall, bald woman with bags under her eyes, and jowls down to Georgia. "You must be Alan," she said, in this very sexy voice, "Come on in."

The room had a sofa bed, hotplate, and shower stall. It was perfect. My stipend was about $350 so it was also affordable. Then Dolores said, "I must apologize for my appearance." I froze, feeling like a fly in a spider's web, unwilling to jiggle the silk. After an extended silence, I admitted an impression that she might have a hangover. This proved acceptable. She explained that her hair had started to turn grey, and she had turned to chemistry. Although an overnight dab of colorant on her forearm had produced no response, full-scale application to her head brought on a severe allergic reaction.

With eyesight and liver affected, she had been hospitalized. Fearing that more would be absorbed in any attempt to wash it out, her doctors shaved her head. I had called just as she had recovered enough to come home.

I moved in, with my few clothes. There was a supermarket close by, and canned peaches were amazingly cheap, so I found myself financially sound right from the start.

Since Dolores' eyesight was permanently impaired, she even offered to sell me her 1957 Buick. Consequently, I was also mobile.

I found this regrettable, however, for that sailboat of a car frightened me. I expected to join the striped bass in the bay every time I started down the San Francisco hills.

After about three years, I decided to live on the other side of the bay, near the Berkeley campus. Needing more reliable transportation for the commute across the Bay Bridge, I sold the Buick and bought a new Yamaha motorcycle, Thor. I found a basement apartment with a view of Alcatraz Island and the sunsets behind the Golden Gate Bridge. Since I didn't expect that I would ever be able to afford such a beautiful view later in life, I drank it in appreciatively.

Social Life

Social events lubricated with California wines brought a cosmopolitan collection of graduate students together. Those from India assumed camaraderie on my part, which was pleasantly confusing. It seemed to derive from the commonality of cricket and tea in the old colonial days. While I had a vivid sense of the effect of Irish history on my family, I had no relatives who had fought in India or administered the Raj. Indeed, the only resonance I could find was Rudyard Kipling. The significance of his poetry is, for me, its vivid undercutting of bigotry. Kipling used the slang of the day to gain the unsuspecting sympathy of his audience, and then, with a drawbridge unfolded into the empathetic hemisphere, he finishes off a poem with an invitation to epiphany: *You're a better man than I am, Gunga Din!*

There was a contrasting difficulty with Chinese students who were uniformly eager to know what I thought about the Opium Wars and the Boxer Rebellion; they were disappointed to learn that I had never heard of either.

When the tale was told, that Britain (through the East India Company) had addicted nearly half the Chinese population to opium, and then refused to stop the extremely lucrative trade, for it was bringing copious supplies of silver into the English treasury, I was incredulous to the point of denial, but it proved to be true.

The first Opium War (1834-1843) began after the Empress of China had personally written to Queen Victoria, pointing out the harm that opium was causing, and citing the strict prohibitions on the trade imposed by Victoria herself within England, Ireland, and Scotland. The new Queen, Victoria, ignored the Empress' complaint and generated a blarneywoman counterclaim: that the Empress' attempts to stop the trade in China had caused destruction of England's private property! For powerful property owners, property destruction is more ethically troubling than human sacrifice. This perverse ranking is a long-running thread in human history. To gain restitution for said damage, the British Queen deployed her large British-Indian army against the Chinese Empress, an action that included dispatch of warships to the coastal towns, with much property damage. But hey, that was their own fault. Right?

Though God should surely have been on the other side, British military might prevailed, and the Empress was forced to pay reparations. China was even forced to cede Hong Kong to British control for 100 years. The final Boxer Rebellion was the peasant's spasm of disgust that the Qing dynasty had failed to protect the nation against all these obvious foreign exploitations. Unable to repel the foreign powers, the peasantry was successful in ending the dynasty. That political shift ushered in modern China, with an understandable chip on its shoulder against the West.

English schooling had left me with an assumption that opium dens were a Chinese invention. My native culture had even turned the self-effacing term "worthy oriental gentleman," into an all purpose acronym epithet: virtually any foreigner could safely be called a wog.

If the West could ever enlarge its context understanding to fully appreciate the misery that its ancestors inflicted upon foreign lands, and the consequences that extend into the present time, then present day diplomacy might become more successful. But diplomats alone will not do the trick; they will need to be representing nations full of correspondingly enlightened citizens.

The Summer of Love

UCSF is a few blocks from the Haight-Ashbury district, and 1967 was the Summer of Love. Curiosity about other people's sex lives is a primate fascination, and the sexual experimentation of the day is accordingly overplayed. The experience that I treasure more was the beaming goodwill that resonated from strangers, the sudden amplification of courtesy, eye contact, and a smile for a passerby. There was widespread social optimism, which extended to a belief that the population at large could surely be persuaded that the Vietnam War was a horrible mistake. Although that summer didn't last long, the fact that it happened at all, that so many took on the persona of Bonobo humans, is worth noting as a primary piece of neurobiological data: our sense of enjoyment and satisfaction responds to optimism and broad expectations of social rationality.

The compulsory joke about how everyone was too stoned to really remember the sixties (a one liner assertion) obscures the latter point. I had no reserves and no place to crash; I couldn't risk a bad reaction and so I didn't take anything. Though this was unusual, I was happy to be sharing peace, harmony and optimism for the future.

After the excessive innocence of the hippies passed, an ethos of random acts of kindness remained pervasive. My motorcycle, blew a tire one day, after I had crossed the Bay Bridge, but before the off-ramp to the city below. I was stuck—up in the air and on the edge of the freeway, wondering what to do.

A Harley Davidson driver with a sidecar pulled up in front of me. I ran up, explained my problem, thanked him for stopping, and told him I didn't think he could help. "But the sidecar's empty!" he said. He had just delivered architectural blueprints, "and the back flap reaches to the ground; we can just push your bike into the sidecar." This we did, quite easily, and then chained up the back-flap. "Now hop on your bike," he said. As we joined the traffic stream, a police car approached from behind. I could see the puzzlement as the officer tried to figure out what the ticket would be for. But then we reached the off ramp and left him for the city below. My newfound friend drove me all the way to the Medical Center, along Haight Street with its legendary residents. They waved at the Yamaha piggybacked on a Harley. I tried to wave back the way the way I'd seen the Queen do it. Once unloaded, I again thanked and wondered how I could repay him. "Oh, just pass it on," he said. I've been trying.

Neuroscience

In the 1960s, many loyalists asserted that President Lyndon Johnson knew more about the "real" reason for the Vietnam War than he could safely state. Those commentators implied that a President should be supported anyway. I realize now that we all lacked a frame of reference that George Lakoff has described in: *The Political Mind: Why You Can't Make Sense of Twentieth-Century Politics with an Eighteenth Century Mind* (2008). The big misassumption is that the same input data ought to lead to the same conclusion in all minds, i.e., the same rational conclusion. Political discussions don't actually work that way, and we need to understand why—we will get back to that later.

This was the immediate backdrop against which I decided to study the brain.

I told myself that I would be doing so in the hopes of eventually understanding the deep roots of dysfunctional politics. I didn't realize then how much I was a product of my family environment, and that those memories represented a still deeper motivation.

That a twenty-four year old graduate student would set out to understand dysfunctional politics might seem grandiose—but I was not the only one who thought that way. Consequently, now that we are forty years older, it has sometimes been disappointing to look at what has been offered as neuroscience-assisted insight into human politics. But the winds of change are blowing. Laurence Shlain, in *The Goddess and the Alphabet* (1998) provided an early blast. I can also feel an eventual whirlwind in the works of Jost et al., (2003) on the personality forces that drive political conservatism. Domke and Coe (2007) have usefully described of the rhetoric of political manipulation, and a very recent release from Iain McGilchrist: *The Master and his Emissary* (2011) also fulfills my hopes from long ago.

What the reader and I are doing herein, however, has a different cast. I am eschewing the standard reductionist brief for a hypothesis in favor of a personal account of how my ideas unfolded and survived crosschecks, and it is time to return to that thread. We need to go back to the old days when the society for neuroscience had about three thousand members, and I was one of the complete neophytes. I had found that The Langley Porter Neuropsychiatric Institute was associated with UCSF, and that I could do thesis work in that environment. George Ellman, in its biochemistry lab agreed to shepherd my efforts.

The intellectual environment of neuropsychiatry sublimely matched my curiosities. Now to figure it all out!

Karl Pribram

In the mid sixties, comprehension of brain function seemed held back by specialist jargon. In response, an interdisciplinary training program at Langley Porter was encouraging various neurospecialists to learn to talk to each other. Over the years, initial enthusiasm wilted under this scornful assertion: *interdisciplinarians know too much about too many things, and not enough about anything.* (Verbal scorn is one of the most durable of left hemispheric behaviors.) The first few centuries of science had been appropriately devoted to descriptions of smaller and smaller components, but the goal of ultimate broad understanding will require interdisciplinary synthesis. Furthermore, if one cares about the human enterprise, it needs to be done before the whole evolutionary drama implodes.

Karl Pribram, a neurophysiologist at Stanford, gave one of the first guest lectures that I attended. He was sparking interest in the possibility that some form of holography might create the massive memory stores we have in our brain tissue (Pribram, 1971). Toward the end of his talk he set up the question: *Who watches what the eyes are seeing? Where is the little man inside the head, and to whom does he report?* He insisted that there was no such little man.

Karl's certainty that a search for an ultimate cause was wrongheaded bugged me, and I challenged it. He insisted that 'I' was no sort of objective thing; that 'I' was probably a process in my brain tissue, not a little man, not a thing at all. He emphasized that it was important to grasp this because it changes one's interpretive stance. I didn't grasp it then as I do now. I went home and tried to design a brain that would have an ultimate criterion event.

I came up with the notion of a Great Pumpkin Neuron that would shrink with pleasure and swell with pain, rendering primordial judgment on all that one did.

For five minutes, I thought neuroscience was going to be a piece of cake. Then I saw that I had proposed a redundant piece of circuitry that could not have been encouraged by evolution. As the idea fell apart in my own hands, I started to see what Karl had been talking about. When a biological process calls me up, "I" happen. I am a process associated with a phenomenon called self-consciousness. As we have noted earlier, consciousness is a narrow, 40-bit per second bandwidth of information surfing on a million-fold greater wave of sensory information flowing into the brain. Nevertheless, "I-ness" illuminate aspects of reality that have survival value, or the phenomenon would not have evolved. Like a light from a hot tungsten filament, my brain creates "me" whenever it seems appropriate. It usually turns me on in the morning, and switches me off at night. You've probably noticed something similar.

Another useful view emphasizes that brain tissue is anatomically designed to support looping pathways of electrical signal traffic. That 'reentrant anatomy' as it is called, encourages the following metaphor: An inward spiral of thought activity drives more neurons within the brain, ultimately creating the subjective sense of, "This is me thinking." An outward spiral of thought activity gets all the way out of the brain and drives muscles, thereby generating what behaviorists call objective behavior. From this perspective, there is no Descartian separation of "mind and matter;" the difference is whether electrical information lands on a set of neurons (producing thinking) or on a set of muscles (producing behavior).

The mind versus matter argument has thus been an example of category fabrication by verbal language. The corresponding reality is actually seamless.

I've noted it before, but the language primacy delusion is worth repeated debunking: the effort to reduce reality to words, essential for manipulating details, inevitably gives rise to logical absurdities.

124

The false division between mind and matter has been a particularly deep absurdity. Remembering previous options of lipstick, toothpaste, and the war in Vietnam, I licked my intellectual wounds with salty pleasure. I could hardly believe my luck. If I could get past the silly stuff, I might have the chance to bloody myself on the cutting edge of human understanding.

The Bohemian Grove

With my Liverpudlian brogue serving as a Beatle-assisted asset, I was welcomed into an era of increasing student involvement in campus decision-making. One day, while acting as a functionary of the Graduate Student Association, I received a call from Chauncey Leake, a vigorous white-haired professor emeritus of the pharmacology persuasion. He had a deep interest in philosophy and a corresponding proposal to make to the graduate students. He limited his initial description to the simple observation that those of us expecting to be awarded Doctor of Philosophy degrees had never actually taken a course in philosophy, had we?

I was asked to come out to *The Bohemian Grove*, and talk it over with him. I asked my thesis advisor what this strangely named place was. George spluttered something like, "Why on earth would you want to know?" I learned that it is an exclusive Bay Area club, in the deep redwoods along the Russian River in Northern California.

The original artistic spirit of its bohemian founders became compromised when funds fell short. As a result, great wealth gradually became a necessary criterion of membership. Indeed, the whole enterprise got stood on its head: wealthy capitalists and politicians had become the backbone of *The Bohemian Grove*. It must be said, however, that they have continued to be vigorous supporters of Bay Area cultural life, which ranks high on the list of all centers of cultural life.

This would be no visit to a smelly toilet to watch cartoons on a whitewashed wall. I began to understand George's surprise.

I was wearing a bright orange crash helmet as I drove Thor up to the gates. A uniformed guard was also surprised to learn that I was on Dr. Leake's guest list, but he graciously let me inside, took my helmet, and called for a honking great yellow bus, which then carried my personal self a full fifty yards to the foot of a wooden staircase. This architectural feat climbed into the trees and out of sight. It was another of those "America is amazing" days.

After receiving the attendant's salute, up I went, eventually joining the rest of the party somewhere near the third star from the right. Hors d'oeuvres were served along with California wines, and a fine time was had by all.

Then Chauncey led us to view the grotto, where stage performances were the highlight of the annual gatherings. Towering redwood tree trunks backed the stage and supported whatever wiring was required to manipulate the stage sets; the seats were carved into downed redwoods. Chauncey described how the stage play was commissioned exclusively for the Grove. As part of the contract, and no matter what its artistic merit, it was only ever to be performed once—by Broadway professionals—and only for the benefit of Bohemian Club members. This description of what I considered to be pathological elitism left me ready to throw up. Nevertheless, Chauncey seemed like a nice chap, and his notion that we ought to get a dose of philosophy seemed rational.

On our way back down the stairway, to the domain of the big yellow bus, we learned that Richard Nixon had immediately preceded us in his pre-campaign meanderings. He had apparently been underwhelming. Knowing Nixon as well as we now do the real mystery is why his mother was ever impressed enough with him to assist his presidential ambitions. I wish I were just joking. As we know to our national sorrow, *The Bohemian Club's* lack of enthusiasm proved no barrier to his ascent to power.

I was apparently being lobbied by Chauncey, and had no trouble pushing the idea of an elective philosophy course at the next board meeting of the Graduate Students' Association. I was even hoping we might use the series to raise questions about the war in Vietnam. That proved to be too optimistic.

Epilogue

In Chapter 7, I reported a personal difficulty with context revision when I encountered unexpected sexual preferences. Five and a half years of living in San Francisco left me with a more enlightened attitude. In the end, all we have to treasure each other for is personal uniqueness.

Rigid context assumptions may grow worse as we grow older. This seems to be recognized in aphorisms such as: *It is hard to teach old dogs new tricks*. In monkey populations, the youngsters seem most adaptable. While the older females retain much of that flexibility, the older males are a recalcitrant lot. Testosterone to blame?

Chapter 9
Singing for the Fire

Musical Enthusiasms

As a result of the last eight chapters, I hope that you can recognize occasions in your life when a context revision enlarged comprehension. Each time that happened, self-consciousness in the left hemisphere gained from the right hemisphere's expertise. In this chapter, I will describe a difficulty that typically interferes with deliberate attempts to deploy the right hemisphere.

The most obvious hindrance is that the right side of the brain will not answer prose questions with prose answers. For a left-hemispheric self that has defined itself in terms of verbal statements, such resolute muteness seems objectionable. The possibility that the silent right refuses to stoop to the approximations of noun categories because it has a 'better' way of describing reality is insulting to the left's ego, and therefore further infuriating. A common response, even among neuroscientists, is to ignore the right hemisphere, and to quickly discount any claims raised on its behalf by third parties.

During our family immigration, however, when we were in Jacksonville in the mid 50s, I saw Elvis perform on the Ed Sullivan Show. I was blown away, as the saying goes. Like many teenagers of the day, I couldn't get enough of him. Why? The media explanation, lurid sexuality, missed the point; it was a deeper phenomenon. He was singing and moving with a strange freedom that does not come from the self-conscious side of the brain.

When we returned to England, I found another idol: the skiffle musician Lonnie Donegan. He was a Scotsman whose physique permitted no chance of exploiting sexuality. Furthermore, his material was the folk songbook of Leadbelly and Woody Guthrie, e.g., *The Rock Island Line*. He assembled a small backing group best described as a jug band without the jug.

Lonnie sang with frenetic energy and the great benefit of a large, highly resonant nose. None of this, however, explains the effectiveness with which Lonnie entrained the adolescents who would become the English Bands of the 60s: including the Beatles and the Rolling Stones. (Mark Knoppfler of Dire Straits not long ago penned an appropriate lament: *Donegan's Gone.*) How did Lonnie catalyze such a powerful cultural response?

I now believe that the secret lay, not so much in his nose, as in his ability to put aside self-consciousness and allow the right hemisphere to drive his singing and behavior. Audiences respond. But I didn't know that then.

When I reached San Francisco in 1965, Joan Baez and Bob Dylan had taken the folk music revival to another level and coffee houses were filled with imitators. Graduate students couldn't afford San Francisco's tourist oriented entertainment, so folk music became my default, one that I now regard as priceless.

I quickly sensed that performers fell into just two categories. The more common kind transmitted self-consciousness through the microphone, while a precious minority achieved something else— they seemed to channel one into the song narrative. I resonated with either type, but sat through the former hoping that they would hurry up and finish. Calling one type amateur and one professional merely labeled the phenomenon, while explaining nothing. Something more important was involved, but I couldn't put my finger on it.

These venues taught me that Donegan's skiffle music had been drawn from the American country blues. Other white kids, like John Fahey, had experienced similar fascinations, and they had even sought out survivors and relaunched careers from the days of the prewar "race" records. Mississippi John Hurt became one such national treasure; his music turned up in most hippie record collections.

A Meeting at the Crossroads

One day, I saw a name I recognized, Son House, and learned that he was playing that night at a bar not far from where I then lived in El Cerrito. He had been a mentor to the young Robert Johnson, the man who allegedly sold his soul to the devil in exchange for the ability to play the blues. Robert's star burned bright for two years; then he died at the hands of a jealous husband, as Son House had forewarned him. I had no idea what House sounded like, but I went down to the crossroads. I arrived early and sat at the bar along one side of the room.

An elegant, slender, and very black man walked on stage alone, carrying his guitar case. A preacher's black string tie and a pinstriped suit contrasted against a white shirt that suddenly became super white under a discrete blue light. The patrons, busy with their conversations, took no notice. Sitting in the chair before the microphone, House opened the case. Out came a shiny, German silver, resonator guitar. I had never seen one like this. For Preacher House, however, the unusual instrument did nothing to quiet the conversational buzz. It was only when he began laying down his rhythm that tongues froze in mid-waggle, and every patron swung around to focus their eyes upon the man. I had an impression of a fleet of starships locking phasers. A moaning, stabbing voice began to interweave with unique sounds as a metal tube on House's left hand slid up and down the strings.

At the same time, the spidery fingers of his right hand were doing rhythmic karate chops all over the instrument. He was singing blues, of course, but with spiritual fervor, and for me he was a black Homer telling of life in chain gangs, in cotton fields, and in prisons in the Mississippi Delta. Here, in the flesh, was another example of someone who had learned to yield to his right hemisphere— and it was jangling mine in the most powerful musical experience of my life. Some performers use deliberate eye contact to increase the audience member's sense of involvement. The singing preacher was doing something else. He was off in an inner world, but he had brought it above the surface where others could see it.

When he had finished his first number, he pulled a white kerchief from a breast pocket, and mopped his fine-boned forehead. Twenty minutes later, he had evidently given his all. With one last mop, House explained that he didn't care to be idolized after he was gone, preferred to be appreciated right now, and thanked the audience for their attention, which had been completely involuntary! I couldn't afford another guitar, but as Thor and I rode home that night, my synapses were doing all the resonating I could stand. An ambition to replicate a cultural feature of the 1930s, bottleneck guitar playing, was being inscribed among my neurons.

For several years afterward, I considered the experience personal and precious. Consequently, when New York's Stefan Grossman started publishing guides to the country blues, I was surprised to learn that I was one of a legion of overawed witnesses of whatever it was that Son House invoked in audiences.

He must have switched on the mojo and stunned many a white boy into silence. I wondered if he reveled in the power of being able to throw that transformational switch. I doubt that he did. Many a fan, envying the status of the star, fails to realize that with self-indulgence one becomes left hemispheric, and the magic drips away.

The situation reminds me of Lot's poor wife who could not resist looking back on the ruined city of Sodom, and so was turned into a pillar of salt. Self-consciousness can be such a nuisance.

The Problem with Trying

We are exploring a non-intuitive and subtle point about the self-defeating role of 'trying' in human affairs. Trying to be like Elvis, Lonnie, Joan or Bob will only produce obviously effortful imitation—arising in the self-conscious left hemisphere. Although musical legends certainly have inspirations, they don't *try* to be somebody else. Their inspirations merely serve as encouraging keys to an escape from egotistical 'trying' so as to, instead, allow one's soul to unfurl in a public place.

The Party Piece

There was a time, not so very long ago, when most entertainment was local, even to the point of coming from within the family. Funny stories were the everyday form, but at extended Irish gatherings, weddings or wakes, a particular party piece was a social requirement and everyone was expected to have some polished nugget available. Although I had started out strongly with *The Song of the Cornishmen*, adolescence had shifted some mental balance wheels, and the idea of performance had become embarrassing. Part of the tradition, however, was to allow for the intolerably shy individual to face the hearth and sing to the fire. As the chosen performer gained access to deeper and broader meanings, the listeners resonated with the singer's soul; empathy and solidarity followed. This bonding mechanism is not there when all parties sit down to face an electronic display.

In the coffee houses of the 1960s, with no fireplaces, I was nevertheless willing to sit through many awkward performances while waiting for the next authentic to turn up. Back then, I was still puzzled at what those two modes of performance represented.

One seminal clue arrived when David Galin and Robert Ornstein, introduced me to the split-brain experiments of Roger Sperry. These provided direct evidence that the left hemisphere's conscious impression of ourselves leaves out the whole right hemisphere.

I felt the benefit of a strangely relevant dream during these years. I seemed to be standing in the interior of a cornflakes box. The perspective was wide but shallow; it seemed to be lying flat on its face. Scattered among the flakes were free gifts, and some of these were pretty women. As desire welled up, I reached for them, but the box began to rotate. I seemed to be on the axis about which centrifugal force developed—and so the objects of desire were propelled away from me.

The more I hungered, the faster the box spun, and the faster the charmers withdrew. The dream continued long enough for me to appreciate the correlations and to resolve not to 'try' for each delight that came within sight. Instead, I would wait for random events to precipitate a more natural interaction and potential adherence. With that decision, the rotation slowed down—and the gifts came closer.

I have generally thought of this as a non-verbal right hemispheric parable presented for the edification of the egotistical and grasping side of my personality. How does the message connect to singing with soul? If the search for status through singing drives the performer, that goal arises from and seems to activate the left hemisphere— which seems to inhibit the only tissue capable of the desired performance, namely the right brain. For performance purpose, it is therefore preferable to kindle ego-free empathetic thoughts that will *allow* the right hemisphere to take the reins.

A recent reading of Viktor Frankl's, *Man's Search for Meaning* (1952) reminded me of the above dream. Frankl was a survivor of four concentration camps and drew his insights from the behavior of human beings in those extreme circumstances. He developed a belief that "fear brings about that which one is afraid of, and hyper-intention makes impossible what one wishes." Under the heading of Logotherapy, Frankl pioneered a psychiatric technique called 'paradoxical intention' that attempts to counter such effects. Something about my far more benign circumstances, including the importance of centrifuges to a biochemist, of pretty women to a young male and of enthusiasm for the phenomenon of soulful singing had led to a remarkably similar conclusion.

Chapter 10
A Tale of Two Hemispheres

PD<1923

The man who stares at us so intently across the years is Paul Broca. His father had suspected that different parts of the brain might drive different behaviors, but false claims that criminality correlated with palpable bony bumps on the skull substituted the soon discredited pseudoscience of phrenology for that valid initial supposition.

Young Broca became a neurologist and so studied brain tissue directly. Nevertheless, in the early part of his career, he became embroiled in another error. As you can see, he had a large head and this may have tempted him into an effort to demonstrate superior intelligence in the Caucasian race with selective measures of cranial volume. That nonsense has been forgiven, however, for Broca also had a more excellent idea.

He supposed that the region that showed damage after a mutism-inducing stroke might define the part of the brain responsible for speech. Consequently, he sought out the brains of stroke victims who had been rendered mute (the neurologist's term is aphasic) just before they died. With subsequent autopsies, Broca showed that whenever speech had been recently lost there was a hollowing out in the left frontal lobe of the brain's surface. The finding has been extensively confirmed. Although Broca realized that the dysfunctional area was only present in the left hemisphere of right-handed individuals, he seemed to view this east-west asymmetry as a secondary curiosity (Broca 1879). By our time, over a century later, it is the asymmetry that seems most significant.

Robert Ornstein's *The Right Mind* (1997), provides a review of the notion that the two sides of the brain are different that dates all the way back to the Greeks. As Ornstein points out, the ancients thought of more possibilities than they had the means to investigate, and so their assertions may well have been examples of trial-and-error loose associations. Nevertheless, the ubiquitous helmets of those sword-wielding days suggest that the opportunity for neurological observations of lateralized damage may have been high.

Here is what Diocles of Carystus said in the fourth century B.C. "There are two brains in the head, one which gives understanding, and another which provides sense perception. That is to say, the one which is lying on the right side is the one which perceives; with the left one, however, we understand." I am not exactly sure what he meant, either.

There is an ongoing problem in descriptions of brain function: can words describe the situation accurately enough to be truly useful? After all, the obviously different specializations of the hemispheres developed before language; retroactive verbalisms may never be quite up to the task of explaining what transpires within brain tissue.

For example, the two dichotomies we have used so far: verbal vs. silent and self-conscious vs. context-conscious are consistent with modern evidence, but there are others that also work. Science never stops and new data may force endless revision.

The Man with the Single Hemisphere

In about 1820, and so prior to Broca's discoveries, Arthur Wigan participated in an autopsy of a man with whom he had been acquainted, and who had seemed perfectly normal. When his skull was opened, however, Wigan was astonished to see that only one cerebral hemisphere was present.

The immediate implication has been born out by much modern research: the developing brain defers to a primary need to talk, to listen to and respond to others—to behave. If only one hemisphere is present, it will learn to do these things, no matter which side of the head it grew in. The curious cognitive 'normality' of single-hemisphered individuals leaves little clue as to what the silent right hemisphere is 'for' when it is present and appropriately developed.

But Wigan realized that two half-walnuts of cerebral tissue present in most of us might support two minds (Wigan, 1985). Once Broca had shown that those of us with two hemispheres only use one to speak from, a peculiarly uncomfortable question became theoretically relevant: *When we use the words "I, me, mine," is the feeling of consciousness also uniquely one-sided at those times?*

The question seems to have been too radical to consider in Broca's day, and it is often ignored today. However, the march of surgical progress inevitably led to a moment when a conscious patient, with an opened skull, was addressed as follows: "The last piece of your right hemisphere has just been removed. How do you feel?" The answer has always been: "I feel just the same" (Austin et al., 1974). This reply has astonished witnesses such as Nobel Laureates (Roger

Sperry, John Eccles) and philosophers alike (Karl Popper). It forces the conclusion that, after right hemispherectomy, the verbal self is still in the patient's cranium, where it has apparently been oblivious of whatever subjective experience may arise in the right.

Now we are faced with follow-up questions: *What is the counterpoint of "I" consciousness in the nonverbal hemisphere? Does it scream silently when surgical necessity requires its removal?*

The Mystery of the Silent Hemisphere

In a pragmatic response to such puzzles, Roger Sperry decided to cut the anatomical rope bridge of interconnecting fibers that joins the two hemispheres in what would be called split-brain cats and monkeys. Once they were thus isolated from each other, he planned to study their intelligence, one at a time.

Sperry used an opaque sheet along the vertical plane of the nose to ensure that different fields of view were presented to each eye. He was then able to communicate visually with each hemisphere separately. Sperry found that either hemisphere was able to solve simple puzzles quite independently of the other. Surgery and cardboard thereby revealed exactly what Wigan had suspected: the two hemispheres have the necessary circuitry to function as two independent intelligence centers. But what survival value had induced evolution to develop two centers of intelligence?

Epilepsy

An epileptic attack begins when a patch of neurons starts to fire in a tight synchrony that overwhelms the brain's inhibitory dampening. Once out of control, the seizures continue until excitatory nerve endings have run out of readily releasable neurotransmitter. In some instances, a burst of uncontrolled firing on one side is transmitted across the connecting fibers causing a 'mirror focus' in the other hemisphere to also discharge abnormally.

In these cases, what one sees externally is that symmetrical parts of the body begin to twitch.

When persistent epileptic episodes have been traced to multiple sites in the silent side of the brain, it is therapeutically acceptable to remove the whole hemisphere. In contrast, removal of the dominant hemisphere, the home of the verbal self, is a much less attractive option. Two surgeons, Joseph Bogen and Phillip Vogel, wondered if, for such severe cases, it might be better to completely section the fiber bridge, the corpus callosum, without removing any tissue—i.e., similar surgery to that which Sperry had performed on animals. The hope was that epileptic attacks would then be unilateral.

Their first patient was a war veteran, and the surgery markedly reduced seizures—on both sides of his brain! It seemed that a two-hemisphere conversation had been necessary for the initial one-sided twitching. Such set-up conversations were no longer possible after cutting the callosum. The clinical success left a philosophical puzzle, for, just as in hemispherectomy of the right hemisphere, the patient seemed otherwise unaffected; there seemed to be no cognitive downside to the surgery. How could a huge fiber tract between hemispheres be cut with little consequence?

The intense form of the question arises from a hidden misassumption: If evolution found survival value in dual hemispheres, is it reasonable to regard the corpus callosum as a bridge that completely reverses their isolation?

Might the interhemispheric fissure instead have a primary function as a barrier, perhaps to ensure hemispheric independence? If so, the knife stroke's relative lack of effect on normal behavior becomes less surprising. At the same time, the callosum is not a vestigial organ; it is robust and it carries signal traffic. A more nuanced function, probably involving a previously unexpected relationship between the hemispheres may be suspected.

Cooperation/Conflict

For a short time after split-brain surgery, the patients were prone to peculiar behavior. For example, one man, who did not like licorice, watched his left hand retrieve a piece from his wife's supply packet. Though the patient knew he did not want it, the hand placed it in his mouth, and he ate it. (He still did not like it.) (Bogen, 2006.) This falls into the category of things we might not want to know, and Bogen even omitted it from his initial research reports.

But similar things happened to other patients. For a few days after surgery, if the executive verbal hemisphere dropped the reins, so to speak, and failed to exert its potential dominance, the silent hemisphere guided the left hand to express a conflicting motivation. Fortunately, patients soon adjusted, and unity of purpose returned. That eventual result was easier to accept (and report).

The Perfect Experimental Animal

Drs. Bogen and Vogel sent their patient to visit Roger Sperry and his colleagues. I have long imagined that Sperry would have been delighted when a talking primate came into the lab bantering pleasantly, needing no restraints, and not soiling the floor.

In fact, I was wrong. It seems that Sperry was uncertain of how valuable this opportunity might prove to be. There was one obvious difficulty: only one side of the human brain talks. The other side is holding its cards closer to its metaphorical chest, and it was this silent side that seems to have most interested Sperry. Monkeys had required many training sessions with food rewards before each hemisphere learned different responses. For sips of chocolate malt, however, humans learned to perform such tasks immediately.

It was the accelerated pace of experimentation that increased Sperry's enthusiasm. Soon an intriguing behavior was noted. Let's imagine ourselves doing the critical experiment.

We flash a red or green light into one eye at a time, in a random sequence. The actual 'test' is extremely simple: we are going to ask the patient to report what he sees.

An Early Split-Brain Experiment

If a red or a green light is flashed into the eye connected to the talking hemisphere, subjects name the color correctly, 100 per cent of the time. In other words, the talking hemisphere can accurately describe the color categories of sensory input. This was no surprise.

But what if the colored light is shone so that only the prose-silent hemisphere can see it, yet the question is still asked, *"What color do you see?"* The silent hemisphere can hear the question, and understands the names of the colors but, as mentioned in the introduction, like the tar baby, it won't answer questions. Yet the experimental result was interesting, for the left hemisphere (which could also hear the question) answered instead. But only 50 per cent of its responses were correct.

A reflexively admiring interpretation might be that the subject was correct half the time. These scientists, however, noted that 50 per cent correct, when there were only two choices, meant that the subject's left hemispheres were volunteering guesses. The investigators were thus left with a strangely familiar conclusion: *The talking hemisphere will guess when it doesn't have the appropriate facts.*

As I noted earlier, there is something about these sorts of results that we instinctively resist. But if a closer approach to the truth is to make us more free, we must surely meet it halfway. Sperry and his colleagues repeated the experiment, with the same subject, week after week. Why? They wondered if new fibers might leap across the surgical cut, or if a different path to transfer information would develop.

Impulse to Cooperate

One day, it happened. Whenever the left hemisphere made a wrong guess, it immediately corrected itself saying, for example, *"Green....no Red!"* It had become 100 per cent accurate—even without seeing the colors!

Where was the talking hemisphere getting its corrective information? The investigators noticed that erroneous guesses caused a pronounced grimace on the left side of the face. Only after a grimace was the correct color named. When the first guess was accurate, there was no facial contortion and no follow-up change of mind. What do you make of that?

The investigator's conclusions were these: when the left guessed falsely, the right heard the mistake (through its connected ear) and initiated action: it pulled the muscles of the side of the face that it controls. The tension from the resulting grimace was communicated across the midline to the other side of the face, which is sensed by the verbal hemisphere. Fluctuations in muscle tension on the two sides of the face were being generated and received as a signal to correct a wrong guess; a real world, Rube Goldberg strategy!

The Puzzle of Folk Performance

The surgeon Joseph Bogen also participated in developing a procedure whereby only one cerebral hemisphere is anesthetized. Called the Wada test, it now gives surgeons prior knowledge of which hemisphere houses speech and so the verbal self; that confidence helps in planning major brain surgery, (Gordon and Bogen, 1974).

They found that when the silent hemisphere was transiently anesthetized, there was a corresponding loss in the patient's understanding of melody. Although still conscious and able to join in with *Happy Birthday to You,* their "singing" (which could only have been coming from the still awake verbal hemisphere) was <u>tuneless</u>. On the other hand, with only the right hemisphere active, some patients had access to memory of the lyrics, and, from this side, they could even use the vocal cords to sing them <u>tunefully</u>.

When I learned that tuneful singing could be driven from the 'silent' hemisphere, I felt that I suddenly understood why Elvis, Donegan, and all those authentics in coffee-shop folk houses had so impressed me: singing from one's "soul" may represent behavior directly expressed by the tar-baby side of the brain. Embarrassingly self-conscious singing may reflect an inability to escape the status preoccupations of the verbal side.

This hunch was reinforced when I learned that poetic recitation is another of the skills that seem to originate in the right hemisphere. For our preliterate ancestors, historical knowledge depended on oral tradition, often as epic poems. I can readily imagine the Homeric Bard being chosen for an ability to escape self-centered preoccupation and so guide the group's imaginations through tales of heroes and wars that preceded their own short spell of corporeal existence.

When I speak about those campfire gatherings an emotional catch burdens my throat, as a Joycian vision weighs heavily. It is a vision of an endless chain of the dead, who were puzzled in life, and who struggled with the "gift" of self-awareness. I feel their relative ignorance as an Olympian Curse. We humans have traveled a long intellectual journey since then. Although it is unfinished, and deep puzzles remain, I am glad that I was excused duty at the dawn of human perplexity. I now remember those coffeehouse days as the modern equivalent of ancestral campfire gatherings.

Elvis Presley and the Hemispheric Shift

The biographer Elaine Dundy spent time in Tupelo, Elvis Presley's boyhood home, where she learned that Elvis' singing enthralled his classmates, even before his voice broke. At the end of the song (usually *Old Shep*) Elvis was back to being just a poor boy from the wrong side of the railroad tracks.

Elvis' favorite comic book character was Billy Batson, the crippled newsboy who would turn into Captain Marvel when he shouted, *"Shazam!"* In later life, Elvis displayed Captain Marvel's lightning logo on a neck chain, and on the tail of his personal jet, (Dundy, 2004). My interpretation of this biography is that Elvis had discovered that he could use singing to drive the hemispheric switch, both in himself and his audience. In other words, his singing allowed a communal stroll across the corpus callosum, and he had learned the trick long before critics wielding one-liners would decide that the adult Elvis was merely exploiting sexuality.

The notion that a hemispheric switch was required for powerful performance art was intriguing. I suspected that other instances of the shift might be equally powerful. But what might they be? I tried to caste a wide net. I'm embarrassed today by how little I caught.

The next one that I noticed involved diving. On my way up the ladder, my inner dialogue was always full of remonstrances: *You've done it before; you don't need to do it again; what are you trying to prove?* These were apparently fearful left hemispheric loose associations verbalized into sentences. As I left the end of the board, by then committed to the embrace of gravity, another command system seemed to push my 'self' aside. Restating: Freefall seemed to provoke my right hemisphere into grabbing the reins. Vigorously implemented jackknife motions then delivered my material self into the water at a safely steep angle, avoiding a painful belly flop.

'I' could remember very little of the resulting descent, my 'I' didn't seem to have been switched on for that segment of time. Once freefall was ended, however, and my self was supported by water, self-consciousness returned with an internally loud assertion that: *I made it!* As with successful singing, I now think such claims are literally self-delusory bullshit. But my cupboard of examples was still rather bare: singing and diving. Decades passed during which I failed to appreciate that every story I told was unconsciously built around the reflexive pleasures of the context shift, and the impulse to laugh that came with it.

Mental Digestion

The above conceptions conflict with intuitive expectation and this leaves them apt to be forgotten like a nonsensical dream. By posing and answering a pragmatic question I will now try to fit them into a more durable cognitive 'chunk': *How does the executive self consciousness in the left hemisphere get any goddam benefit out of a tar baby hemisphere that studies the context as an almost covert operation? What on earth is the point?*

The following demonstration provides an answer. A saltshaker is placed on a table and those sitting around are asked what they see. Most will say, "A saltshaker." Each individual has thus used speech to self-report their own brain scan of what they 'see' in consciousness—the saltshaker. But reflected light from the tabletop is surely entering the eyes; the table is in no sense 'hidden'; everyone saw it, but few report the fact. Why? The background context had not been classified as a detail that belonged in consciousness.

The simple demonstration makes evident a useful division of labor: the executive left consciously appreciates those few details that might be relevant to impending behavior, for example, shaking salt from the shaker.

At the same time, the right brain maintains broader levels of contextual detail in reserve *on a need-to-know, rather than on a covert basis.*

In other words, researcher's frustration with right hemispheric silence may have obscured the possibility that consciousness needs overload protection. Tabletops require either legs or pedestal for their own support, and the floor, in turn, must support these. The underlying joists are set into the walls of the building that rests on foundations embedded in the earth of the town, county, state, nation, hemisphere and planet. We are more or less aware of all this in our world model, but clarity in consciousness may depend upon keeping it in the background.

Autistic individuals have a well-known need to avoid overstimulation. They rigorously hone their sensory input down to a very narrow spectrum, particularly avoiding eye contact. Perhaps failure of an early sensory or a later cognitive filter has demanded a cumbersome behavioral compensation.

For routine problems of daily living, forty bits per second passing through conscious awareness seems ideal. When circumstances demand, broader considerations from the contextual stores in are added, as needed or, perhaps, as requested.

Jill Bolte Taylor and the Silent Witness

A professional neuroanatomist has single-handedly provided a new chapter in this long-running saga. Jill woke up one morning with a severe headache behind her left eye. (We aren't acquainted; I'm taking the liberty of using her first name because I feel a sympathetic affinity and want to express it.) She was in the midst of a left-hemispheric stroke. A blood vessel associated with the areas that support the self-circuitry was leaking, and the left hemisphere was intermittently going off-line.

When the left side was temporarily functional, she was self-consciously aware that she was in mortal danger and that she should call for help. Each time it shut down, however, she experienced a unique form of consciousness, one that was no longer worried about her 'self' in the same way. This frame of mind presumably arose in her still functional and no-longer inhibited right hemisphere.

Although her language skills were deteriorating, when the left side was intermittently functional, she placed an arduous phone call to her lab. Fortunately her colleagues recognized her garbled tones as a plea for help, and an ambulance was quickly on its way.

What was right hemispheric consciousness like? Her book title says it well: *A Stroke of Insight* (Bolte Taylor, 2008). Jill's sense of personal separation from the rest of creation was diminished, perhaps as the left's enthusiasm for categorization failed to dominate. She reports that boundaries lost their hard-edged quality; reality seemed fluid and she felt that she was continuous with that fluid. In this state, worry simply ceased.

Instead of worrying about what she should be 'trying' to do, the non-self version of Jill Bolte Taylor experienced a deep appreciation of the 'now' and the cosmos, apparently the condition called nirvana. In that state, the cause-and-effect logic of the left hemisphere, and even its obsession with the fate of the self vanished. Jill knew she had been granted a holy grail experience. After eight years of relearning language, she told us all about it.

Meditation techniques have led to similar testimony, claiming that awareness of a larger, more significant reality than that of the fragile self is available through training. In one tradition, repetition of a mantra allows the perceptions of the Silent Witness to become conscious. (Perhaps the meaningless mantra-word sedates or bores the verbal hemisphere so much that it 'allows' the reins to be taken by the right hemisphere, an easier to control and more reversible technique than a stroke.)

As Jill relearned verbal skills, presumably exploiting the neuroplasticity of the tissue around the stroke site, she enabled her left hemisphere to regain dominance. As it did so, it seemed to increasingly disallow or inhibit the right, for she found it became more difficult to experience the simple bliss she had enjoyed earlier. Her conviction of the general validity of the right hemisphere's perspective assisted her, however, in editing and adjusting the self's inner dialogue. In her book, she reports how she deliberately practiced techniques for quieting both self-critical and status-hungry verbalisms, thereby retaining a broader perspective.

Epilogue

There was a crooked man, and he walked a crooked mile,
He found a crooked sixpence, upon a crooked stile.
He bought a crooked cat, which caught a crooked mouse,
And they all lived together, in a crooked little house.

Chapter 11
The Trouble with Icons

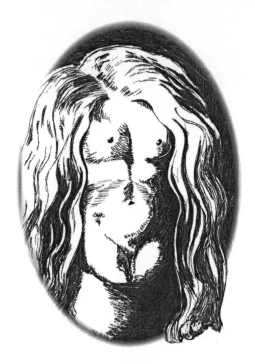

Is the Right Hemisphere Infallible?

The colored light experiment, in the previous chapter, revealed that the verbal hemisphere has a reflex urge to answer when it doesn't have the appropriate data. In the present chapter, we will consider a corresponding vulnerability of the silent hemisphere, and thereby preempt the thought that this tissue might be infallible.

Background information: Both isolated hemispheres recognize that a frontal view of a face is, in fact, a face. The silent hemisphere can do so quickly, and will successfully analyze even fuzzy images. The verbal hemisphere is slower; its preoccupation with details seems to limit its perception of the overall context.

New information: Dahlia Zaidel (1990) found that when side-view-silhouettes were tried with split-brain patients, only the verbal hemisphere knew what they were. Why was the silent side ignorant? A pair of staring eyes is a powerful biological signal, full of survival implications. It seems that the silent hemisphere has found the associated iconic outline useful. Side view silhouettes seem to be of less significance to right-hemisphere-world.

The investigators knew of a curious painting by René Magritte, entitled "The Rape." Although this was a full-face view, it was unusual. As you can see in my drawing at the head of this chapter, a female hairstyle frames a face shape. Magritte, however, had composed the details, the face's features, from nude body parts. The eye, nose and mouth positions were occupied by breasts, navel and pubic mound, respectively.

Although the silent side avoids speech, it can understand language. Consequently, the investigators were able to ask the subjects to point to named face parts. When instructed to point to the "eyes," the silent hemisphere indicated the nipples. When asked to point to the breasts, the hand went below the painting's frame, and so pointed to a region where the chest would fall on the same scale! In contrast, when the verbal hemisphere saw the painting, the patient immediately commented on the anomalous details. The investigators were stunned. The isolated silent hemisphere had recognized the outline of a full face, but it had then filled in expected features—without resolving discrepancies in the details.

This is immediately reminiscent of two issues. One is a simple physiological matter: there is an inevitable blind spot on the retina where the optic nerve enters from behind the eyeball. The void is there as you read, but the empty zone is routinely filled-in using context information from the surround (on the rest of the page), and so we are generally unaware that we have a blind spot.

Conspiracy Theories

A significant proportion of most civic societies today seems convinced that they live inside a massive conspiracy. Disregarding Occam's notion that the simplest explanations are more likely to be accurate, such theorists cite innumerable details that could fit with conspiracy, but which either have simpler alternate explanations, or are plain counterfactual. The certainty they ascribe to the conspiratorial conclusion, however, betrays a deep emotional investment. Anyone challenging such views is immediately assigned the role of a gullible sheep being herded along according to conspiratorial whim.

Since I must assume that all of us have a deep emotional investment in our operating (context) assumptions, I should disclose mine. I am convinced that civilization is currently irrational and that it must change. There are certainly vested interests that will resist such change. To that extent, I feel aligned with conspiracy theorists. I don't, however, see enough intelligence or conscious organization in those vested interests to successfully carry out the supposed conspiracies. Instead, I see a general vulnerability of all sides to reliance on pseudo-logical, one-line simplicities. I am trying to get to the bottom of that quite specific vulnerability before suggesting a response.

Back to Zaidel's work. To confirm the right side's tendency to confuse details in favor of an overarching belief, the investigators turned to simpler line drawings, as shown on the following page. When questioned about the one on the left, the silent hemisphere again pointed to the proper place for facial features, rather than the actual places shown in the drawing. Even in these simplified renditions, the right side's perceptions were still slaved to the overall outline! If they were sure of the big picture—that this was a face—they could not see conflicting details.

When the outline was removed, as shown on the right, the subjects pointed accurately to the jumbled features. In other words, isolated silent hemispheres are quite capable of resolving and recognizing jumbled facial features—but only when they are not surrounded by iconic contexts, such as a facial outline. Iconic cues receive some form of meaning-priority, which can preclude an accurate analysis of details.

Review

The verbal hemisphere has a tendency to shoot from the lip, producing false assertions. The silent hemisphere has gaps in its repertoire of context insights, as well as a tendency to ignore details when a definite context familiarity is present.

This doesn't bode well for the longevity of the human species, and one might even wonder how we have made it this far. Part of the answer is surely that the two hemispheres are normally able to cross check each other. In that sense, behaviors seen in isolated hemispheres of split-brain patients may not be net vulnerabilities.

Nevertheless, honesty and due diligence demands that we pursue these potential vulnerabilities further. For example, are there situations in which overarching context assumptions trick normal humans into accepting or believing inappropriate details?

Do political movements sometimes create overarching assumptions that are difficult to correct by citing countering details?

Bartlett's Phenomenon

In 1932, a professor at Cambridge, England, published a book called *Remembering*. Sir Frederic Bartlett's focus of interest was the party game in which a whispered story is passed around a circle of participants until the initial version can be compared with the hilariously different final version. Bartlett wanted to understand what prevented simple accuracy as a story passes from mind to mind. This was a penetrating choice of experimental question and his results are relevant to the intense political debate that must surely accompany an effort to build a more rational civilization.

In the days before tape recorders, whispers were an impractical sort of experimental data. So Bartlett wrote out the first narrative and gave it to the first person in an experimental chain. The instruction was to read the text (only once, and no further cross-checking was allowed) and then rewrite it for the next person. By so doing, Bartlett generated a set of written accounts that he could peruse at length, searching for the origins of the errors.

It became quickly evident that, in a first round of processing, the early facts were being used to deduce the contextual frame of reference. Only after recognizing the frame do we decorate it with specifics. *And once we think we have recognized the frame, we bias those specifics.*

Bartlett realized that we do not start with an open mind. We bring personally preferred, right hemispheric frames of reference to the task—context understandings that we are already familiar with. In other words, we carry around our own iconic assumptions.

A natural consequence is that we can only attach those new details that will fit. We either ignore, forget, or force-fit details that don't match. By arranging for this tendency to be repeated around a circle of participants, the party game magnifies the effect to the point of bemused laughter. That laughter masks a concern we should all feel.

A major underpinning of civilization is an effort to produce accurate reports, such as peer reviews, court records, minutes of committee meetings, opinion journals, etc. All these are driven by a need to build the consensus needed for society to function efficiently, uncovering mistakes—and there will be mistakes—before too much harm is done. Since 1932, we have not formally recognized how much that effort has been held hostage to hidden differences in context assumptions.

Summary

The silent hemisphere can leap to broad conclusions that are inconsistent with available details. In cooperation with the verbal hemisphere, it can then mangle details to fit the mistaken context assumption.

Epilogue

Those who seem most convinced that every detail confirms conspiracy have a pronounced tendency to assert that 'experts' are in agreement with their thesis. Those 'experts' that I have tracked down turned out to be better described more critically. This leaves me with the impression that the definition of expert status (a status-centric, left hemispheric obsession) may also be held hostage to an ideology of rampant conspiracy.

Chapter 12
Racism: Active and Passive

PD<1923

In 1967, I became president of the Graduate Students Association at UCSF, and membership in various committees came with the position. I had noticed that this sort of thing granted some sort of status and initially enjoyed the experience. However, I quickly concluded that committees involve a conservative bias towards the status quo.

My simplistic analysis included the thought that consensus for change was being delayed because each member sought to present a unique perspective, apparently for ego maintenance. "I disagree, therefore I am," as Descartes might have put it. There is more to it than that, and the Dilbert cartoon strip has become a syndicated exploration of the very many ways that committees subvert long-term idealism. The need for an efficient election of a new Pope provides an example of the countermeasures sometimes introduced. The relevant committee is locked inside the Vatican and its members aren't allowed out until they are finished. (They are even forced to communicate with smoke signals.)

Coming to this cynical view during my year as a functionary, I disentangled from those efforts, and volunteered instead for an assistant editor's position on the campus newspaper, *Synapse*. I dreamed of motivating constituencies that would lock committees in rooms, leaving them to decide their way out.

Beware of what you wish for.

Assassinations

Just five years after John Kennedy was shot down, Martin Luther King met a similar fate. His death, which he had clearly foreseen, became the context for my activism.

Once upon a previous pivotal moment, the national student organization had hired buses to drive volunteers south to register black voters. One bus had been burned in Alabama. The perpetrators disappeared into a community that was still nursing the rationalizations of their ancestors, and who were therefore deeply frightened by a phrase: 'outside agitators'. That fear drove lynching and mutilation. The passion that is required to mutilate someone for disagreeing with you is not exclusively a phenomenon of so-called backward foreign lands.

The Teach In

This time, in 1968, it was suggested that each campus arrange a Teach-In to discuss racist practice in local surroundings.

As we sat down in the Student Union, just a few blocks away from the Haight-Ashbury district, we were a motley crew. None of us had in any sense dropped out, but the weird and wonderful fashions of the time had penetrated the medical center. For medical students, often bearded and wearing sandals, bell-bottom jeans were in vogue. If a tie was worn, its color and size hinted at the LSD experience. Dental students held down the conservative end of the spectrum: they were close-cropped, clean-shaven, and further recognizable by their short white jackets.

I differentiated myself from both by wearing a Canadian sheepherder's fringed leather vest. It extended down in a kind of beavertail over my bum. I learned later that this extension was supposed to be tucked into the trousers, so as to prevent the cold Alberta wind from chilling one's kidneys while one bent over a sheep. Having no intention of getting that intimate with a farm animal, I just let the flap hang out for all the honest world to see.

After the organizers made a general statement, the first meeting of the Teach-In seemed to stall, for there were no designated teachers. Group dynamics eventually squeezed individuals to their feet and I still remember the first soliloquy from a black medical student who described two-culture schizophrenia.

As he left a ghetto environment in the morning, none of his acquaintances would believe that his day would be spent in medical school. As he returned to the 'hood in the evening, he had to shift his perspective back to theirs. Though his remarks were eloquent testimony to the reverberations of centuries of slavery, they did not identify any specific local racism.

In one of the smaller meetings that afternoon, a contingent of dental students seemed to feel that they were on a mission for which the moment had come. They described how the UCSF Dental School had once trained both black and white students, but, fifteen years previously, an end to that practice had coincided with the appointment of a new chairman of the Admissions Committee. This professor would readily offer extended theories of black inferiority to whites, the highlight of which was the alleged racial failure of African peoples to develop writing. He also proclaimed disbelief that black peoples had a culture of any kind, and so felt that they should be forced to conform to white preconceptions. Later on, we will consider the forces that propel racist blarney in some depth. For now, I'll just assert that, to this powerful chairman, these were self-satisfying but false justifications for discriminatory behavior.

The account of a passively institutionalized injustice struck me as exactly the reason we were having a Teach-In. It also provided the opportunity I had sought in volunteering for newspaper work. On the spot, I declared that I was willing to provoke immediate reform by writing up the story for *Synapse.* As had happened occasionally in childhood, I felt certain that a small part of the future was now in my hands. I could not resort to hitting him over the head with a bicycle, of course, but I was confident that a straightforward account would arouse public opinion, forcing committees and executives to act.

Knowing that the facts had to be bulletproof, I made my offer contingent upon one or more of the dental students giving me a formal affidavit to the effect that they had directly heard these bogus notions. To my astonishment, none of them would; they ruefully explained that dental training involved daily checks of work on outpatients. Any bias against them could result in their grades staying low, so that graduation would be delayed and accumulation of debt would extend indefinitely.

Winkling Bigotry

I rode my motorbike home in deep frustration. A mild earthquake threw a sway in the Bay Bridge, which shifted lanes beneath us. I adjusted, still more focused on winkling out a bigot than on tectonic plates. As often happens with frustrating mental puzzles, I just needed to relax and fall asleep. I awoke with the realization that I only needed one affidavit—my own. If the descriptions had been accurate, a simple conversation should provide the evidence.

In postwar England, the water had not been fluoridated, and one's original teeth rarely lasted into retirement years. By providing myself as a patient to learn on, I was getting short-term amalgams replaced by durable gold work. The student who was doing this arranged that my next appointment would occur when Dr. Illigitime would be on the floor, checking the work.

This was easily done, and once the rubber dam was off, I found the loose association that would elicit comments about black people. "Well, their problem is just that they aren't as well developed as white people. After all, they didn't even think of writing all that time they were in Africa."

As he spilled out his racist tract, exactly as reported, I felt like Brer Fox as he watched Brer Rabbit hitting the tar baby. I had a clear perception of the consequences I was determined to impose upon him. This required a certain bloody-mindedness, a deep conviction that stupidly self-serving rationalizations from many members of the leadership class unforgivably blight the lives of those below.

Perhaps John Kennedy's death, the daily body counts of friend and foe in Vietnam, and now Dr. King's assassination gave me the necessary determination. However, everybody who read newspapers was informed of the context we were all living in. Some additional force was operating in either my bosom or my cerebral cortex.

My dental-student collaborators made sure that I knew the broader circumstances of their lives. Over half of the dental school faculty was ex-military; they were apparently a cheap source of experienced staff—who brought their culture of father-discipline with them. This was further manifested in a rumor that there existed a secret list of students thought to have smoked marijuana. It was feared that their graduation would be impeded.

Since my persona would evoke the Antichrist to this crowd, I decided to blow their cover preemptively, forcing a discussion of secret lists, and hopefully extinguishing any such extracurricular enthusiasms. A question here: Was I being arrogant?

As I check that editorial, forty years later, it looks surprisingly innocuous. I framed the argument carefully, acknowledging everyone's right to have controversial opinions and to express them freely, but then I forcefully questioned whether the rest of the community should passively sit by while those opinions were turned into de facto admissions policy. I did not advocate that the offender lose his job, but only that he be removed from the chairmanship of the Admissions Committee. It seemed so simple.

A nagging thought bothered me: Dave Bomar, who was the senior editor of *Synapse*, was a dental student. How would he react? I turned the article over to him and waited. He replied quickly. "Since this is the last issue of *Synapse* for the year, would you mind if I showed your piece to some people who might wish to respond right away, before the summer vacation starts?"

Today I would firmly say, "No, they've had fifteen years. It's my turn." Being young, and open to honest argument, I said, "Sure!"

The next week was hell. It was clear that I had found the edge of institutional propriety and, for some people, gone too far. As I walked down corridors, office doors seemed to open by hydraulic magic; curled fingers beckoned me inside.

Sometimes it was tenured professors who offered me anonymous encouragement. I was relieved at this but had to control the feeling that their status was supposed to make my efforts unnecessary.

The Good Guy

There were others. One professor pointed out, very sweetly at first, that he sympathized with most of what I had to say and that he had been fighting the good fight from within. He just wanted me to know that he himself had been recruited from the military and that my comments were damning him along with the real bad guys. Wouldn't I just revise that aspect of the article?

When I pointed out that the existence of allegedly progressive exceptions like him did not undermine the validity of my generalization, his face darkened and his eyes grew so narrow, a length of dental floss would have blinded him. He ushered me out of his office with barely controlled fury. I thought to myself: This is one of the good guys?

Dean Ben Pavone of the dental school called me, asking for a meeting. He began by making it clear that he was not going to ask me to withdraw the article. (I suspected that the Free Speech movement had cleared my path in this regard.) He said that he had asked for our meeting so that he could acknowledge that he knew these problems were real. He just wanted me to know what he was going to do about them so that I could decide if the article really needed to appear. He described an ad hoc admissions committee that he had created, and which would carry out the real admissions process in the future. The standing committee would continue to exist but would have no further power.

I instinctively liked this man and believed he was sincere in planning change, but I couldn't understand why such a charade was necessary. I asked why he didn't simply remove an obviously inappropriate individual on a committee? I wanted a description of his frame of reference—his context understanding, as I would call it now. He sighed and said, a little condescendingly, "The university system just doesn't work like that."

The thought swelled again: *I'm a nonentity, but by simply publishing the facts, I can make the goddam system behave like that! Why the hell can't you?* I wish I had said that to him: we might have had a more extended and deeper conversation. Unfortunately, I held my tongue.

The challenge to institutional practice felt like facing a cornered animal, one that pretended to accept the principles of open debate claimed by the wider culture, but which could not, in fact, tolerate a spotlight on its bad behavior. Now Dean Pavone had persuaded me that forces were in motion that would seek to keep the institution away from the edge of this racial cliff. As we parted, I was feeling the relief of knowing that I had decided to back off. Fortunately, I did not try to ingratiate myself by saying so without further thought.

I walked more lightly down the granite steps onto Parnassus Street. But another finger beckoned me toward the ineffective cover of a large potted plant. Here we go again. This time, it was one of the original student whistle-blowers, who quickly whispered that a rumor was going around that I had been recruited by the Communists to foment trouble on campus, and that plans were afoot to have me deported. (I no longer know this man's name; if he should read this: Thank you, from years later—you made the necessary difference.)

Gripped by icy calmness, I now knew that professional survival meant that the whole community should know what might motivate such a knife in the back. The article had to appear.

I called the senior editor, ready to threaten that I would hand out copies on the Parnassus Steps and invite the attention of the *San Francisco Chronicle*, if necessary.

No such words were needed; he accepted my decision.

Six pages of rebuttal accompanied my two-page article. There was no factual challenge to any of my allegations and no new information. The substance of those six pages was that I didn't know the whole story, and "one shouldn't write such inflammatory articles without knowing the whole story." Nevertheless, three months later, the Admissions Committee had a new Chairman. It seemed to me that I had moved the system.

I hung up my leather vest, seeking anonymity, and concentrated on finishing a Ph.D. thesis. I felt that I had been true to myself, had acted appropriately, and had done enough. I also realized that my involvement in any other cause would be perceived as a kiss of unwelcome radicalism.

I remained deeply puzzled that in this land of free speech and tenured university professors, it had been left to an immigrant from Liverpool's dockland to put principles into practice. I was similarly puzzled at the firmness of my own conviction. Acting as an impromptu social blacksmith, I had hammered a small piece of dysfunction into better shape. Was this the way in which progress has always happened? It had been a lonely exercise, and now I was afraid for my own future. Would the passionate iconoclast reemerge unpredictably? Would I be an army of one, when it did? I was darkly certain that society was riddled with like problems. I felt sure that I would both uncover more and respond. But how long could I dodge, duck, and deflect fury? A sad, compelling certainty developed that the standard path through a professional career would not be possible for me. At some point, I thought, I will slip-up, and a cabal of authoritarians will declare me a menace to 'their' society. Although I did manage to develop an academic career, and gain tenure, it was from this time forward that I was consciously looking for a Plan B, and a more independent place to stand.

Epilogue

This story has a surprising resolution. To tell it, I have to jump forward some thirty years, to a time when Plan B had been put into effect, and I was living in the Great Northwest. One morning my friend Ed Stiles was hosting a visitor, Hercules Morphopoulos, and the three of us met in Friday Harbor's Front Street Café. I don't tell the above tale as a normal part of my back-story, because there isn't much to laugh at. When I learned that Hercules was himself a dentist from San Francisco, however, it began to unwind.

It soon turned out that Herc knew more about these events than I did, for he had then been working in the Dental School! Now privy to the rest of the story, I learned that I had not, in fact, been waging lone warfare. Dr. Bob Brigante, Herc's mentor in dentistry, together with a few other professors and a Black Caucus association (consisting of some doctors and many nonprofessionals, including janitors) had been mobilizing as part of an unpaid educational opportunity committee. The Dean's longtime executive secretary, Maybeth Green Monte, whom I must have met when I spoke with the Dean, was the real heroine for she had played a sustained and vital role in soothing conservative fears over proposed changes.

While Martin Luther King's assassination, the Teach-In, and my resulting editorial squeak had collectively broken the back of conservative resistance, the groundwork, a more sustained effort, had been laid long before.

The result: Two black students had been enrolled for the following year. Subsequent efforts led to a fully multiculturalized student body, and encouraged similar efforts in the medical school.

The deposed committee chairman, interacting with actual black students, with their culture and personality in his face, had lost control of himself. His bigotry had become overt, and the Dean had been forced to terminate his appointment. At that point in Herc's story, I slapped my thighs, spilling coffee all over the table.

Who the hell do I think I am? I am sure that my sudden involvement in this drama led to that exasperated question in many minds. When one exercises the rights of man, however, there is no obligation to explain oneself. Nevertheless, I also wondered about the source of my conviction. Was it pugilistic ancestry, or *The Song of the Cornishmen?* I doubted that any special intelligence had been involved; it felt more like simple honesty.

The real events in this chapter conform to a continuing problem: some of us evidently harbor an authoritarian and antidemocratic world-view that manages to coexist with the flatly contradictory principles upon which America is founded. Why are their views so rigidly held? Who the hell do they think they are? Though we aren't ready to tie it all together yet, the bare bones of a possible answer are already latent, as follows.

In the previous chapter we saw how, in a split-brain patient, an iconic assumption based on a facial outline led to mistaken detail analysis. In this chapter, we explored the case of an authoritarian conservative whose brain was presumably intact but whose context understanding (of racial superiority) was so iconic that he mangled the detail analysis. So a refined question might be, why was he unable to correct his apparently long-held view of the world?

Although fear is an obvious probability, it is not a useful answer on its own. We need insight into what part of our evolutionary inheritance drives such fears; how they are instilled, generation-by-generation; why they might drive lynching and mutilation; and, above all, how the cycle might be broken.

Chapter 13
Of Cambridge and Electric Fish

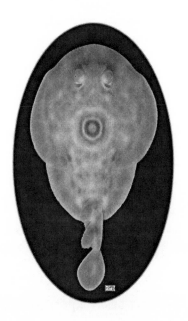

My Ph.D. Thesis

I usually explain my Ph.D. thesis in terms of fried eggs. Egg white becomes opaque in a frying pan because of the rearrangement of sulfhydryl linkages in its major protein, ovalbumin. There had been a report that sulfhydryl chemistry in the brain was altered by neuronal activity. Since my thesis advisor, George Ellman, had invented a better reagent for measuring such changes, we set out to see if thinking was anything like frying eggs.

I used electroconvulsive shock to produce seizures in rat brains. Although seizures are pathological versions of 'thinking resonances' they should have had measurable effects. Better measurements, however, revealed that the various forms of sulfhydryl reactivity that I defined remained unchanged.

The original claim was close to fraudulent. As I wrote up my thesis, I found out that it had even been retracted. A positive discovery would have been a better start to an academic career. To make matters worse, many biochemists were then in long supply and driving taxis, for example.

Fortunately, I learned of a program to bring English scientists back to the motherland. This would give me a small salary for a year, which I could earn by doing research in an established laboratory. I applied for and received that funding, and began my Samurai science period, five years of postdoctoral wandering, searching for access to a tenure track position somewhere, somehow. My first position was in Victor Whittaker's lab in Cambridge.

An iconic theme in approaching a scientific puzzle is to find and study the simplest possible example of the relevant phenomenon. Since the Renaissance, laboratories around the world had recognized the virtues of electric fish for the study of biological electricity. By our time, it was known that briefs squirts of acetylcholine turn the organ into an equally brief forty-volt battery that drove considerable current through seawater and so stunned the fish's natural prey. What was still unclear was how the nerve terminals managed the squirting process.

We will need to use the word acetylcholine again, so let's demystify it. Acetic acid is generally known as vinegar. Choline is a sort of ammonia. Your nerves drive your muscles into contraction by spraying ammoniated vinegar onto them. This has overtones of the tin man clanking around Oz, which is one reason professionals prefer the latinized jargon. Although I believe the image of us clanking around the cosmos is usefully self-defecating, I won't press the point and will use the highfalutin' term.

The electric fish evolved its electric organ from chest muscles, but it had to adapt to a penalty in the process.

The resulting tissue (a kidney shaped mass on either side of the gill slits) is not as flexible as muscle and, of course won't contract. Consequently, these fish are poor swimmers.

When I once saw newborn Torpedoes swimming, I was strongly reminded of Flash Gordon's spaceships from the old time Saturday matinees. Although there was lots of personality in the effort, something seemed not quite right. They make up for their deficit by hunting in the dark, using an exquisite sensitivity to electrical fields that can detect their prey's muscle movements (e.g. the heartbeat). Once they locate a victim, rays fold their wings forward and turn up the voltage. (*How do you like yours done?*)

Electric organ tissue an ideal starting point for applying the grind-and-find procedures of biochemistry to the question of the acetylcholine spraying mechanism.

Cambridge University

In a tearoom very different from the canteen of my factory days, I learned of an acquired deficit in my verbal skills. To explain this, I should first note that the traditions of politeness developed in the overcrowded English culture ensure that outright physical violence is left to the ruffian class. Status contests are normally limited to verbal swordplay. In the generally optimistic Californian society of the sixties, however, the put-down instinct had been put aside. Like the man hitting his head against a brick wall, I noticed the relief when I found that I could stop.

Consequently, on my return to an English center of intellectual elitism, I was out of practice. The reality hit home while I listened to a description of someone's research. I needed clarification, and said, "I'm sorry, I don't understand that point, could you explain?" The response was: "Are you serious?" After a long pause, my colleague tailored an explanation suitable for a retarded eggplant.

One Wednesday afternoon I further discovered that the status-blind directness that had been validated in California, and had suited my personality, still mangled the sensitivities of my native countrymen.

The head of the biochemistry department, an elderly professor, introduced the guest speaker, Professor Thoenen, from Sweden, and then sat down in the front row. Thoenen's English followed a pattern of using all his available vocabulary to encircle a concept for which he lacked the precise word—one got the point with a crossword-puzzle-flush of pleasure. He told us of giving a drug to alcoholic humans, a drug that blocked alcohol metabolism. He said that alcoholics liked to participate.

Critical minds respond to such nuances with questions. Did the alcoholics come for the free drinks? Or did the inhibition of metabolism make the buzz last longer? Is it the alcohol itself that has the "feel-good" addictive properties, or is it the metabolite produced in the liver that affects mood? Thoenen seemed to be in a position to answer these significant questions, and with my family background, I was interested. Yet he ignored them, moving on to what apparently interested him more. As he did so, the white-thatched, professorial head of the department chairman tilted to one side.

By the end of the lecture, these early questions were still unanswered, and I watched from the fourth row as our esteemed leader was nudged awake. He collected himself, then dug around for the question he had composed before napping: "Tell me, Professor Thoenen... did the subjects maintain... a state of inebriation... throughout the duration... of the experiment?" He must have loved the intellectuality of the rolling multisyllabic words but they were inappropriately complex for a native speaker of another language, and the query went right over Thoenen's head; he answered with what sounded like a boilerplate response to a different question!

The forms of give and take had been observed and an apparently satisfied silence followed. Wondering how people who could play verbal slash and burn all day long, could simultaneously be passive in the face of this Alice in Wonderland farce, I raised my hand. Thoenen nodded in my direction. *"I'm sorry. I'm afraid I didn't understand. Did the subjects behave drunk?"* In tight synchrony, the three rows of people in front of me visibly winced. In an instant epiphany, I knew that I did not belong in England. Three rows of people evidently felt the same way.

What had I done? It was probably an escalating set of three things. First, despite my early elocution lessons, I was still enunciating with broad Liverpudlian vowel sounds; the "U" in my "drunk" evoked a more Anglo-Saxon degeneracy than the genteel pronunciation. Second, by leaving the "enly" off "drunk," I was putting the need for clarity ahead of grammatical correctness. Finally, by alluding to the incomprehensibility of the previous exchange, I was refusing to respect authority for its own sake, and this may have been my biggest faux pas. I had again played the Emperor's-got-no-clothes card. But I was not a little child. I should have known better. As the reader might suspect, I did know better. Just as I had feared, my inner Celt had reared up and swung his rhetorical shillelagh.

As clarity became primary, we all learned that the Swedish government had put restrictions on how much alcohol research workers could give subjects. That amount had been insufficient to produce overt drunkenness, so the points of interest had been beyond the reach of the experiments. It was nice to know.

On the Virtues of Live Fish

The Thoenen incident had happened during my first month at Cambridge, and it made my survival in science still more dependent upon exciting research achievements.

But a major practical problem arose: Victor Whittaker's lab was then reliant on frozen electric fish brought back across the Atlantic from summer stints at Wood's Hole in Massachusetts. After we had ground up the tissue and centrifuged the mess, the test tubes containing the synaptic vesicles were supposed to be detected by way of their acetylcholine content. It turned out, however, that frozen tissue loses all its acetylcholine. This meant that we couldn't be certain of having isolated the vesicles. I protested that this undercut our potential severely. Victor attempted to exert dominance. He pointed out that my existing record as a scientist was unimpressive, and that he would be unable to recommend me for a future position—if I didn't get some results from the frozen fish.

There was no Rumpelstilskin to turn to, and I had already found that no glory could come from half-assed research, so I decided to stand my ground. There was a simple way to do it. I went down to the library and wrote a letter to Victor (so that I could not be misrepresented later), in which I reminded him that I had brought my own salary. I included a point blank statement that I refused to work on dead fish.

This tense situation was resolved in a group meeting, which had been arranged to take place in my absence. When the full group also expressed their misgivings, (probably more politely), Victor made a trip to Paris to see a former colleague, Maurice Israël. Maurice gracefully put Victor in touch with the director of the French Marine Station at Arcachon, who arranged to send us live fish. In so doing, Professor Israël saved my career. (I was delighted to meet Maurice years later, and thank him personally.)

Victor's recent research papers had emphasized the impressive purity with which he could isolate synaptic vesicles from the electric organ, using a new technology called zonal centrifugation. Initial analyses had suggested that the vesicles enclosed a more complex chemical than acetylcholine. Unfortunately, it seemed to be a rare type of nucleotide and it was resisting identification.

In this situation there was an obvious and simple analytical approach: to take an ultraviolet spectrum of the freshly purified vesicles, in the hopes that a characteristic nucleotide spectrum would show up. This may not have seemed obvious earlier because of all the effort that was being focused on the centrifugation technology but, once that was working, spectroscopy was the next logical thing to do. I made it my first priority and immediately found the characteristic spectrum of adenosine was present. The most common form is adenosine triphosphate which drives most energy-requiring activities in biology. Fireflies use it to generate light, and humans can use dried firefly abdomens to detect and measure ATP by counting the light flashes. Mike Dowdall and myself quickly did the assay, and thus learned that cholinergic synaptic vesicles have a double stuffing of acetylcholine and ATP. The paper came out in *Biochemical Journal* (Dowdall et al., 1974). It now seems that ATP is present in all synaptic vesicles and many other types of secretory granules. I had been part of a major discovery.

Human relations being very much as Machiavelli described them, I still had a problem. I had stood up to Victor's bombast, provoked a major change in plan, and then made an obvious contribution to his research program. Machiavelli warned that the junior party in such relationships was vulnerable to a complex response. I knew that I had permanently antagonized Victor. Right or wrong had nothing to do with it. That's where punting on the Cam comes in.

Punting on the Cam

Cambridge's college buildings have well maintained lawns that run down to the Cam, a decorously winding river. If you can find a maiden, punting on the Cam makes a very romantic date.

The punt itself is a flat-bottomed, elongated rectangle with a place to stand on either end, so that you can place a pole against the bottom and push while moving hand over hand up the pole—thus delivering momentum through one's torso to the punt.

When the top end of the pole is sunk below shoulder level, it's time to concentrate on getting it back. This would be fairly easy if the bottom was rocky, but it's not: the bottom end is stuck in the mud. If retrieval involved amounts of pulling equivalent to the aforementioned pushing, one would merely oscillate in an elongated spot, wondering how more skillful practitioners actually go somewhere. (While laughing gaily with gorgeous women.)

The trick is to twist the pole, which releases it. Now you should again go hand over hand in the other direction, pulling the shaft out of the water as high as you can get it. Did I mention the bridges? That's why it's called Cam-bridge. They are delicious stone confections of the past and they occur wherever a road passes over the river.

For pedestrians who lean over the bridge wall to gawk at the punters, it is wonderful fun to see some poor novice who has just figured out how to get his pole out of the mud, and has it balanced high in the air ready for another push—only to turn forward and discover a low-slung stone bridge rushing toward him.

There being insufficient clearance for a fully erect punting pole to pass under the arch, the Hobbesian choice is whether 'tis better to abandon the pole and stay in the craft, or to hold onto the pole, allowing the punt to sail off.

On any summer afternoon, there is always some poor soul who, in the extremity of the moment, assigns the pole the greater importance. To be clinging to a vertical pole planted in the middle of a river is surely an original experience in anyone's life. But it doesn't last long. Inevitable imbalances lever said pole out of the mud—putting nearly everyone in a good mood.

And how did I get to be so familiar with all this, you might ask? You can ask as much as you like, but I have to get back to more important matters.

Having learned the hard way, I was asked one morning to do the punting for a visitor from New Orleans. Paul Guth was a professor at Tulane and an admirer of Victor's work. His lab had been trying to determine if the first chemical step in mammalian hearing involves an acetylcholine squirt. His test animal was the guinea pig. After playing music to the live pig, he was collecting the fluid in the cochlea to see if any acetylcholine had been released. The news that the transmitter was co-packaged with ATP must have intrigued him by raising the possibility that music would also drive ATP squirts into the ear fluid.

It is hard to exaggerate the pleasure of being alive while poling along the River Cam discussing such interesting issues. (I'm leaving out all the relevance to medical considerations such as hearing loss—those form a major section in grant proposals and are perfectly valid, but the average basic scientist is appropriately motivated by broader, more philosophically based curiosity.)

As my year in Cambridge drew to a close, Victor did not offer to fund me for a further year, and I did not expect him to. Nor was it obvious how I might arrange for another position back in America. I began to think that I would have to return to the ocean waves. At that point, I received a telegram from Paul Guth: he had just been awarded a multiyear NIH grant and wanted to incorporate electric fish work into his lab at Tulane Medical Center. Since he was also going on sabbatical for a year, he wondered if I would I consider running the lab for the first year of his new funding? There are times when it seems that there must be a divine providence, and this was one of them. Yet the earliest behavioral strategy in evolution is simply: behave.

If you do something, anything, chance will then unfold danger and opportunity. While punting on the Cam, Paul had found a postdoc and didn't even ask for a recommendation from Victor. I accepted Paul's offer.

Just before I left Cambridge, a German postdoc, Herbert Zimmermann, joined the group and he began experiments in which he would stimulate the ray's brain to control the electric organ discharge. Apart from general knowledge of the ray as an animal model, the most useful things I learned in Cambridge were from Herbert. Our brief encounter had the curious character of an atomic collision in which direction and momentum were exchanged. Until that time, I had considered myself to be a biochemist, and Herbert had emphasized electron microscopy.

For the rest of my research career, however, I was doing electron microscopy, while Herbert emphasized biochemical details of the vesicular content. This is the reason one wanders around doing postdocs, and there would be nothing to regret about it, except for Victor's antipathy to yours truly, as described in the following chapters.

Epilogue

Acetylcholine is only one of many neurotransmitters. Dopamine (DA) is another, well known for its presence in the pleasure circuits of the brain. For our later purposes, gamma amino butyric acid (GABA) is worth mention here. GABA is the major inhibitory transmitter in the mammalian brain. Moreover, inhibition seems so essential that GABA is the most abundant of all the neurotransmitters we have. Now why would that be?

One way to find out is to block the effect of GABA—something that can be done with the drug, strychnine.

The poisoned animal becomes tense and stationary, as if on edge. It seems to have a premonition that the slightest stimulus—a handclap, for example—will drive it into massive convulsions.

In most quadrupeds, the anti-gravity muscles are the strongest and, as the strychnine-poisoned and convulsing animal dies, their influence gradually predominates. The result is the cartoonist's death pose: limbs straight out.

Conclusion: The brain's design seems to have demanded the general dampening that massive deployment of GABA permits.

There is also a more psychological form of dampening that is of specific interest to us. Evolution seems to have found survival value in gently restraining the naturally self-centered, egotistical behavior of the executive left hemisphere. We experience the effect as a seeming whisper of conscience. Said whisper may arise in the more spiritual, non-self and unselfish hemisphere.

Nevertheless, the speech-amplified phenomenon of bullshit seems to regularly outmaneuver the whisperings of conscience. We still have a ways to go.

Chapter 14
The Trouble with Levees

Great Stone Buildings

I had occasionally hurried through London in my traveling shoes, too broke to tarry. Though still poor, I now felt that I could stay with a friend for a few days, and look around. On those previous occasions, I had been as awed as Dick Whittington's cat. But discussions with foreign graduate students in another country had raised some niggling questions about colonialism. I remembered that central London had burned to the ground in 1666. The great stone buildings in their place must have cost a pretty penny. Had it all come from the natural resources of a small island?

I was no longer ignorant of the doings of the British East India Company and the powers it had been granted to govern and exploit India; of how that monopoly had spilled over into an opium trade that poured silver into the British treasury. I remembered the fatted pigs from Ireland. I still didn't know about Cecil Rhode's deployment of machine guns against spears in diamond-rich Rhodesia, nor of the atrocities sanctioned by other crowned heads of Europe. But I had developed a general opinion that exploitative profiteering had been camouflaged by missionary work. So how much of London's architectural greatness represented decent bronze, and how much was twisted spelter? Who could unravel the accounts?

I have only recently learned that I could well have asked the same questions about Liverpool. As I have now noted in the introduction to this book, my home city's development can be clearly traced to the stimulus of the Slave Trade, but nobody mentioned this to me when I was growing up. In fact, I was twenty-two before I learned that some cellar environments in downtown Liverpool still hosted manacles fitted in the brickwork of the walls—to hold slaves awaiting dispatch on a different ship from the one they came in on. I had often been to *The Cavern*, where the Beatles had played, with no inkling of who might have similarly packed those dank spaces in the century before.

Conclusion? In 1971, I couldn't produce any. I didn't have a grand scheme, a ledger of profit and loss that could be attributed either to high versus mean intent, or to the machinations of one hemisphere versus the other.

I have, however, learned that those personalities who view the past with patriotic nostalgia become quite strident in denying that indefensible practices tarnish national history. So I'll just note that I now looked on great stone buildings with a doubtful eye.

New Orleans

I arrived back in New York, grateful for another crack at a professional career. Since the milk and money rumor remained ephemeral, I traveled the rest of the way to New Orleans on a completely full Greyhound bus.

A woman beside me was going to Alabama with a baby on her knee. I was astonished at the vocabulary and erudition of the child; he reminded me of the Little Lord Jesus, and all like that. When mother and son disembarked, I felt compelled to comment upon his profundity. "He's a dwarf," she whispered to me, evidently having paid for one fare, and taken full advantage of the children-travel-free policy. I was amazedly filing this story in my memory banks when we stopped a little farther down a country road. The door opened, and the whole bus developed a pronounced sag towards said door— as in a four-wheeled genuflection.

Oh good grief, I thought, nobody could be that big. Besides, the only empty seat was the one just vacated by the mother and the Holy Child, right beside my personal self. I saw him soon enough and he really was a sight. I exhaled and squeezed up against the window. Nevertheless, as his shoulder swung around it sent my glasses flying, and I was pressed like a daisy.

After discovering a body configuration that permitted respiration, I learned the story of Big John. Unable to read or write, he had put his bulk to use working railroad lines. He was a modern-day, one-man, gandy-dancer. (Sections of railroad lines were once eased into alignment by a team of men with levers. They used a work song to coordinate their efforts as they "danced" the line into its proper place.) The work tended to be increasingly outside of towns and, when he could, he rode the train, which would have had a stiffer suspension.

As often as not, however, he grabbed a Greyhound and then strode through the brush to where the train tracks needed his tender mercies.

Arriving in New Orleans, I carried a small suitcase in one hand, a plywood guitar case in the other. The humid city felt like an engine room, and I was highly motivated to find an air-conditioned apartment, which I did, near the tramline that ran down Saint Charles Avenue. Arriving at the lab next morning, in the Tulane Medical center, I was introduced to Marion Stockwell, Paul's chief technician and major practitioner of the art of acetylcholine bioassay. Dr. Charles Norris was part of the group; he was doing the work that would soon show that Beethoven caused transmitter squirts in guinea-pig ears (see Guth and Norris, 1993).

Also in the group was Tim Bohan, now M.D., Ph.D., who was developing ideas for a thesis with Paul. Tim had a collaboration going with a colleague in the Anatomy Department, Dr. Terry Williams, where he was learning the skills needed in electron microscopy.

Since the show was running fine under its own steam, I just set up the fish tanks and went fishing. At a marine station in Biloxi, Mississippi, the director had learned that, at certain times of year, some deep hollows filled with another member of the electric ray family, one that Harry Grundfest had introduced into the Wood's Hole research community in the 1950s. I set out with the director early one morning, eager for another adventure on the water. But the small boat was inappropriate to the challenge of a high wind that day, so we trawled the bayous for shrimp, taking home enough for massive gumbos. A little later, we found a supplier in Texas, who mailed us our first fish. I quickly confirmed that the isolated synaptic vesicles contained both acetylcholine and ATP, and noticed that snake venom, which degrades lipid membranes, released the neurotransmitter a lot faster than the ATP.

181

The Scientific Method

Practitioners of religion and politics have long deployed dogmatic assertions about how our reality came about, and or, how human affairs ought to be arranged. Their assertions are invariably tied to the trappings of status, and we are status-sensitive creatures.

The scientific method begins with an acknowledgment of the original blankness of the canvas of comprehension. In other words, the starting point is acknowledged ignorance. The next step is to imagine competing explanations for any phenomenon of interest. By methodically testing each explanation, science chisels away its own errors, gradually revealing more robust explanations that may therefore be closer to the truth. But science is never finally and absolutely 'right.'

Nevertheless, this professional devotion to error detection pisses off nearly all non-scientists for lay folk often perceive error accusations in terms of status sonar. They then readily believe that they also heard the scientist claim to be *absolutely* 'right' and therefore, smarter than they are—at which point one-liner psychology takes over from any reasoned discussion. Since the layman rarely distinguishes trivial 'rightness'—over the time of day, for example, from conceptual accuracy, scientists are vulnerable to accusations of arrogant faith in their personal infallibility.

I am trying to work against this by showing how much the experience of error is responsible for eventually pushing analysis of any issue towards greater accuracy. Here comes the instance that made my scientific career.

Synaptic Vesicles

At Tulane, in our first attempt to view the Torpedo ray's synaptic vesicles using an electron microscope, we managed to destroy them all.

It happened because rays are cartilaginous, and the saltiness of their body fluids differs considerably from bony fishes, as well as from mammals. Unfortunately, we had used standard, too-dilute mammalian fixation solutions. The vesicles had promptly swollen and burst.

As we discussed the debacle afterwards, I remembered Herbert Zimmermann mentioning that a famous electron microscopist at Harvard, had recommended putting calcium salts in fixation solutions. The claim was that it would better preserve membrane details.

Looking up the reference, I found that Karnovsky had added a rather small amount of calcium. Since we needed to make up for a great deal of missing tonicity, I chose a concentration that was an order of magnitude higher. On this basis, Victor would later suggest that I had taken proprietary information from his lab; I'll explain that shortly.

In an earlier chapter, I mentioned the knight with the tiniest lance in Christendom—Sir Bernard Katz. (If you just had any loose associations, I encourage you to note whose memory banks they came out of!) Sir Bernard had once used his microelectrode to impale a frog's leg muscle, and record the electrical response to small quantities of transmitter released from its still-attached nerve ending. He had shown that the secretion process, whatever it was, completely depended upon the presence of calcium ions in the bathing solution (Katz and Miledi, 1965). He suggested, however, that those synaptic vesicles full of neurotransmitter inside nerve endings might have a binding site for calcium—which might also serve to zipper them to the inner face of the nerve ending, in preparation for the squirting event. He included a speculative line drawing.

Unknown to us, several groups were adding calcium to fixation solutions in the hope of seeing such calcium binding sites.

Nobody else, however, was adding such huge amounts.

When we got to the electron microscope for the second round, so to speak, we found, as shown in the diagram, that we had now preserved the vesicles. Moreover, there was a distinct black spot of bound calcium in nearly every one. Curiously, our images looked like they had been copied from Sir Bernard's diagram, for they included membrane-attached vesicles! And so we were the first, albeit accidental, discoverers of the anticipated result.

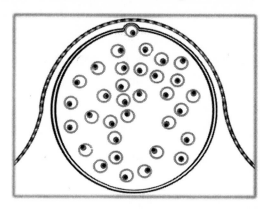

This was also worth an immediate letter to *Nature*. I like to think they asked Sir Bernard to review it, for it was promptly accepted, (Bohan, et al. 1973). That coup set up a series of lengthier papers, also in prestigious journals. I felt that Victor, back in Cambridge, had feared that this might happen. His fatherly advice had been to quit using the electric ray as a system because, if I did, I would always be viewed in his shadow (as one of "Victor's boys").

I had been appalled at an underlying not-so-subtle implication that he had intellectual rights to the electric fish model; that he was supposed to discover how neurosecretion worked.

Now that I had stumbled upon an intriguing phenomenon of which to pursue the significance, I had every intention of pressing ahead on my own.

As Paul Guth returned from sabbatical, I was invited to move with Terry Williams, who had just accepted the anatomy department chairmanship at the University of Iowa in Iowa City. This meant that I could maintain access to electron microscopes and continue explorations already begun. I accepted that offer, too.

The Klingon Harp

Before I left New Orleans, my interest in music and resonator guitars took another turn in the form of a curious dream. It was prompted by the graveyard at the bottom of the road, which featured coffins held above ground in masonry mausoleums. This seemed an irrational practice until I learned the history behind it.

When the levee breaks, the Mississippi River returns to its old floodplain. Flooded cemetery grounds frequently soften enough for buoyant coffins to erupt, sometimes floating their occupants back to favorite haunts on Canal Street. Wisely, the city fathers had decided to do all burials within concrete enclosures.

During the day, I had escorted visitors, including two children, through the local cemetery. We found that each mausoleum had a broken glass plate at one end. Bones and scalps could be seen inside, evidently worked over by animals. The children were thrilled.

In the dream, later that same day, I had been enjoying a late evening in the French Quarter when I heard an intriguing guitar sound and traced it to a street musician. As I approached him, I saw that he was an elderly black man, which deepened my interest. It was around midnight and he had just closed his guitar case, evidently calling it quits.

Then he set off, bent over and limping—in the direction of my apartment. I followed behind, hiding as seemed necessary behind a magnolia tree or in a driveway. The graveyard hove into view, at which point my small hairs began to lift.

He turned at the cemetery gates, and looked around. I stepped back. When I looked again, he was gone—but the gate was open. I ran up in time to see a trailing heel passing down a row on the left-hand side. I tiptoed in and looked around that corner on my hands and knees.

The guitar case lay on the gravel in front of one of the mausoleums. Although the bluesman himself had vanished, there was smoke curling out of the broken view plate. I stepped quietly up to the case, opened it, and beheld the most lust-inspiring instrument I could imagine. Moreover, there was a folded note with it. I picked it up and later remembered the unfolding process with clarity. It read:

If you can play me, you can take me;
But if you can't — best leave me be.

Such moments in dreams seem to be direct indications of an independent intelligence separate from self. The message was well tailored to my inner insecurities; I knew that I couldn't play the way I thought I should. Within the dream, I also felt the flash of fear—fear that my potential for thievery had been foreseen by an entity with broader wisdom. So "who" composed this insightful scenario?

Yet now I hungered for the guitar. Should I snatch it up and run for the graveyard gate? Would the curse extend beyond and follow me through life, a hellhound on my trail? The dilemma proved excruciating: I woke up.

I now see dreams as trial-and-error resonances focused on patterns rather than logical linear analyses. Various aspects from my personal narrative had elicited mutual resonance and were being randomly woven into a semi-plausible fantasy. They are freeform riffs on the available content of our brains: neuronal jazz.

This particular dream was also unusual in the sense that I can trace its further effect on real life.

No, I didn't start stealing guitars, but I was eventually spurred into trying to create a real-world version of that voodoo instrument. The final version has ten strings, six of which are in three pairs. One of each pair is tuned to be slightly flat, while the other is sharp by a corresponding amount (this is only practical with modern chromatic tuners). The brain averages the tones from each pair, hearing a single appropriately tuned note, but the beat effect of their mutual interference is also audible. The guitar, known locally as the Klingon Harp, has attended many dinner parties. Played with a brass version of the bottleneck, it facilitates my escape from self-consciousness, allowing me to sing from the context dictated by the apparent mood of the original songwriter.

For those reading this text as an ebook from the ibookstore, that technology allowed me to design a corresponding musical and poetical experience, one that will have novelty for many readers. Music, poetry and newness all elicit right hemispheric attention. Consequently, this may provide a direct demonstration of what I can only indirectly indicate with prose sentences.

The experience will only be completely novel, and so most dominantly right hemispheric, once. Consequently, I encourage you to take responsibility at your end: arrange for the first listening to be in as relaxed and as unstressed a circumstance as your life allows. (Perhaps just before going to bed.) Think of the Greeks around a campfire listening to Homer tell stories, for the song is a tale of America's past. It was written by my poet collaborator, Tom Phillips, whose own volume (*Tug Boat Man*) is being published at the same time as this one.

Chapter 15
Iowa City Igor

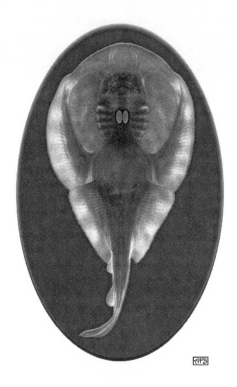

The Sweet Smell of Survival

Just before I drove up to Iowa, the *Nature* letter appeared in which we showed how synaptic vesicles could be seen to bind calcium. I immediately received an aggrieved letter from Victor, who had himself moved to the Max Plank Institute in Göttingen, Germany. He claimed that he had been, "just about to study this," and had been advised that I had learned of his intentions while still at Cambridge. This implied that I was a dastardly idea-thief, someone who might prevent him from getting a Nobel Prize.

Victor's reaction was childish, but I was angry, particularly since I could imagine his unwritten allegations were even more scurrilous. Nevertheless, I had a head start on an interesting phenomenon, so I went tripping off down the yellow brick road to Iowa.

Soon after arriving, I had a delightful meeting with Rodolfo Llinás, an electrophysiologist with a philosopher's breadth of interest in the mechanisms of mind. Rodolfo had recently been in Woods Hole, where he had been following the research efforts of Alberto Politoff, Steven Rose, and George Pappas. These three had been adding calcium to fixation solutions for frog neuromuscular junctions—in the hopes of seeing calcium binding to synaptic vesicles!

The discovery that we had already been there, done that and established priority was probably no more welcome to Pappas' group than it had been to Victor. This time no nastiness was involved, and I developed friendly relations with George and his colleagues, who published their results the following year, (Politoff et al., 1974).

Rodolfo's excitement over the potential implications of our findings broadened my understanding, and in Iowa, I had my own laboratory. Electric rays from the Texas gulf coast were flown in, and I soon had interesting experiments underway. I began to feel as though I were doing more than merely surviving as a scientist: my confidence was growing.

In return for ready access to electron microscopes, I was required to help teach gross anatomy to dental students. Since I had never learned this subject, it would take some effort. I shared the student's interest in the brain and the pelvis, and resolved to do my best with everything in between.

Nerve Terminals in the Electric Fish

When I was not teaching, I was studying the nerve endings of overstimulated electric organs.

I had learned from Herbert Zimmermann that one could tie the mouth of an anesthetized electric ray to a tube and pump oxygenated seawater over the gills, to keep the fish alive. By cutting the nerves to one organ it could be completely prevented from firing, and by placing stimulating electrodes in the brain, the other organ could be fired at a chosen rate. The two sides could then be compared by whatever techniques one had available.

I knew that Herbert, Mike, and Victor were planning to study the effects of stimulation on the biochemistry of vesicles isolated in sugar gradients. In Iowa, however, I could see two different reasons for examining the electron microscopic appearance of nerve terminals stimulated to exhaustion.

In 1973, two scientists then at NIH, John Heuser and Tom Reese, had shown that prolonged stimulation of frog nerves caused synaptic vesicles to become fused with the outer membrane of the terminal, increasing its total surface area and changing its shape (Heuser and Reese, 1973). I found the potential for a shape change intriguing. *Do nerve endings in our brains wriggle and contort while we think?*

The other reason was this: Sir Bernard's sketch had implied that synaptic vesicles, as they became added to the outer surface of a nerve terminal, should transfer their calcium-binding capacity. In that case, if we drove most or all of the vesicles into the terminal membrane (by stimulating the tissue to exhaustion) that membrane should take on the appearance of a string of black pearls.

There was another possibility, however, one that I privately thought more likely. I suspected that calcium was binding to all that ATP that Mike and I had shown to be present. In this case, the discharge of collapsing synaptic vesicles should spray both acetylcholine and ATP, so the capacity to bind calcium would have been lost. The figure on the next page illustrates the two possible outcomes.

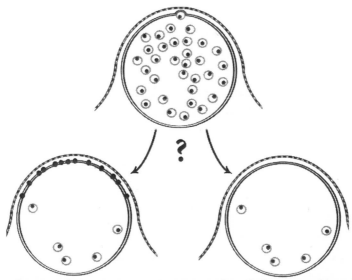

At the top, the resting nerve terminal and the vesicular calcium binding capacity are represented. The lower left shows what we imagined would happen if the calcium binding capacity was transferable. On the lower right, the alternate outcome is shown, i.e., the expected appearance if the calcium binding capacity vanished during vesicular discharge. This is a classical sort of scientific question: two different possibilities and an analysis that should clarify which one takes place. For those readers who care to go with me, I can take you a little deeper into the art.

Although one can set up an experiment that may seem to have just two possible outcomes, the fabric of external reality usually contributes complexities. As one gains experience in science, adolescent obsessions with being right or wrong give way to an appreciation of hypotheses themselves, for these form the caterpillar tracks upon which the whole enterprise moves forward.

There are a seemingly large number of us whose egos prevent a firm grip on this abstraction. Editors of popular media encourage this by continually casting science stories as battles between right and wrong individuals.

The public media enterprise would do civilization a greater service by teaching how science uses errors in its march towards broader perspectives.

Why do Seizures Elevate Mood?

While the fate of black spots was a clean question, and certainly justified the experiment, I had also recognized a loose resonance with my earlier thesis work, in which I had learned that electroconvulsive shock therapy really does relieve depression in many patients. To this day, nobody understands why.

After Heuser and Reese's results, it was natural to wonder if seizures might have their therapeutic effect as a result of shape changes in nerve terminals. I was in a position to create overstimulated nerve terminals and then stay up all night studying them, for as long as I cared to do so. This was precious to me in itself, and I hoped to thereby reap tangible intellectual benefit from my disappointing Ph.D. thesis.

Note that I had no idea why a shape change might benefit mood. Neither was I ready to ask the question in brain tissue, for I didn't know exactly what to look for. Nevertheless, there was a possibility that the extremely simple anatomy of the electric organ might help me define a shape change driven by hyperactivity. If so, it might prove to be a stepping-stone, one that might 'someday' be the basis of something practical to ask about shock therapy. In these two motives, I again see the balance of the two sides of the brain.

What happened to the black spots? was a sneer-resistant detail-concern suited to the linear logic of the left hemisphere. On the other side of the brain was the nebulous hope of a new context for understanding electroconvulsive shock therapy. Although a reductionist might disparage the notion, it was the broad, right hemispheric vista that got me out of bed in the morning.

192

(An aside: I was still living from paycheck to paycheck, but I was well aware that it was a tremendous privilege to be able to think such thoughts and to do such work. The educational opportunities I had encountered were certainly driven by mixed motives, but I had never needed to grow so much as a potato on my own behalf. Since childhood, my family and civilized society had provided basic biological needs while I made the intellectual efforts that had brought me to a point where I could spend my days indulging simple curiosity in the hope that it would be useful.)

The diagram below shows what happened. As expected, most of the vesicles disappeared. But they did not transfer their calcium binding capacity to the terminal membrane. So that outcome was clear.

Yet Mother Nature had also done her thing, for something odd had happened: there were occasional small nerve terminals inside

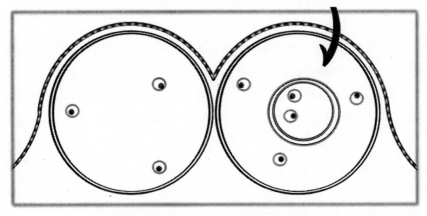

larger terminals (arrow). This was normally the wrong place for nerve terminals to turn up; I knew of no other examples. What did it mean? Here is a context-challenge rumination you might like to try: what three-dimensional architecture would explain the above two-dimensional cross section?

I've just illustrated how science-speak confuses everyone. Let me try again: How might one nerve terminal get inside another?

If you can spread your context net wide enough, that should be a no-brainer. While the wheels of your unconscious are thus turning, I'll digress with another tale from Alan Boyne's schooldays.

The Brethren Deploy the Mental Forceps

Among small English boys of bygone years, breast enthusiasms were supposed to end with weaning, genital curiosity was taboo, and sex education unthinkable. In my case, the Folies Bergère in Paris and then the packet of photographs in Fort Myers had provided the default mental focus for the hormonal changes of puberty.

My family was still living in Florida when the categorical event happened: some instinctive circuitry had induced me to develop a romantic relationship with the mattress. With no prior warning, the first ejaculation could only be interpreted as a heart attack. I fell asleep convinced I was a goner, and awoke surprised to be able to get out of bed. I re-approached the same territory the following night, and decided that this was a wonderful experience.

After all their coy allusions, the Christian Brothers fraternity might be surprised at this. We had probably intuited, correctly, that they had some sort of problem of their own, and we just tuned out when they replayed their homilies. There came the day in adolescence, however, when the Brethren decided that it was their solemn duty to instill durable resonance between shame and puberty, and they announced a three-day retreat at school. Unknown to us, we were about to be inoculated against our natural instincts.

Regular classes were suspended while we prayed, meditated, listened to sermons, and went to Mass continuously. The big gun was an outside priest. Since we had not had a chance to develop a nickname for him, his penguinhood was undiminished, and this may have lent gravitas to the sermon on the third day. He began abstractly, with the usual stuff about lust, whatever that was.

This exhortation passed harmlessly in and out. Still intent on adjusting our neurons, however, he went on to specifics. We were told that thoughts about what was hidden beneath women's underwear were "forbidden by God," under pain of mortal sin and eternal damnation. If that wasn't bad enough, a new word, "maarsturbation," was firmly connected with mortal sin and the aforementioned penalty. (I'm using the Irish spelling.) Like a matador delivering the final thrust, he alluded to the self-stimulations that constituted this maarsturbation business. My affection for the mattress fitted into the framework of evil; one by one, we must all have felt the penny drop, realizing that we had been sinning mortally every night for some time now.

A commotion suddenly developed in a pew behind me: one of my classmates had rolled over into the aisle, and then began to thrash around, clearly possessed. It was pretty obvious that he had done it once too often, and must now pay the piper. We curled our heads still further down, deep in shame.

From the corner of a horrified eye, I saw Dusty Coleman still more mysteriously wrap a handkerchief around his own thumb and stick it in the kid's mouth. The miscreant was then dragged from the chapel.

Although we had never seen him having a seizure before, we later learned that our classmate was epileptic. The good Father had shored up his own rationalizations, his own vocation in life, by driving an undeserving, innocent adolescent into an epileptic attack. I doubt that he felt any shame at all.

That seizure was merely the visible effect. There were also long-lasting guilt trips. The potential relief was confession (in which a priest absolved one from the dire repercussion of eternal damnation), or to get married. We were too young for the latter; it is too normal a behavior to stop, and so we just had to live with obvious destiny.

I only felt safe after Saturday morning confession. By Sunday morning I was back among the legions of the hell bound. Do such teachings reach the level of arrogant emotional abuse?

Porno-Electron Microscopy

If you figured out what the nerve terminals were doing, you probably realized that the above story was not a digression, after all. The most likely interpretation of the strange cross sections was that interdigitations had developed between abutted nerve terminal swellings.

Once again, you can't use a phrase like that and expect people to know what you are talking about, so I also drew the diagram below. It was very difficult for it meant overcoming years of repression about these things. As you can see, there is no way to look at it without thinking about those things of which we have been told not to think. What would Pongo have said? Could I really publish this?

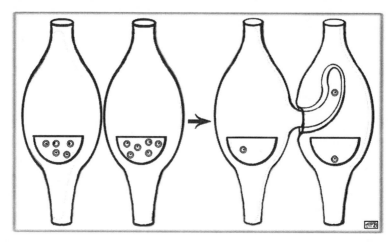

The Journal of Cell Biology readily accepted the paper, (Boyne et al., 1975), but a reviewer suggested that my hypothetical drawing should be dropped. If you think about it, it isn't all that hypothetical. I suspected that the reviewer was concerned with propriety—a status obsession that often reduces creativity, as well as insight.

Nonetheless, I readily acceded, for I needed to produce as many publications as possible to gain tenure and, if pornographic speculations were unacceptable, then by golly, I would collect pornographic data. But to clarify the third dimension, I would need to cut long strings long strings of ultrathin cross sections sliced from a tissue that was now embedded in hard plastic.

The chain of loose associations that connects scientists in their endeavors is endlessly interesting. A Brazilian scientist, Eduardo De Robertis, had participated in the original discovery of synaptic vesicles, and had also developed the diamond knives with which one might cut serial thin sections for electron microscopy. Furthermore, as an already established scientist, he had once visited Victor's lab when Victor was unknown.

Some months after his visit, De Robertis published electron micrographs of homogenized rat brains showing how a population of pinched-off and resealed nerve endings ended up in a unique zone on sugar gradients. Although Victor had been first to report mysterious 'bound' acetylcholine on such gradients, he was not personally capable of electron microscopy, and so couldn't determine what he had. But, once again, he had felt that the rest of the world should wait for him to arrange for electron microscopy and so maximize his credit. De Robertis, instead, followed the published recipe for Victor's sugar gradients, did the microscopy, and produced pictorial results himself.

Although Victor later won the naming wars, calling the structures *synaptosomes*, De Robertis' subsequent Nobel Prize (for his prior years of work on the synapse) did nothing positive for Victor's mood. Now another generation of idea thief, yours truly, was using a prior thief's invention on Victor's electric organ. Ain't life awful?

The cut sections lack any inherent mechanical strength, and so they are allowed to float off the knife-edge onto the surface of water in a trough, held there by surface tension. It requires calm patience to cut long strings and keep them in register. I eventually got the hang of it, and prepared the necessary serial micrographs.

Computer software programs that might reconstruct 3-D models were in their infancy and, instead, I used simple carpentry. I projected the magnified images onto plywood of the appropriate thickness, and drew the necessary outlines with a pencil. After bandsawing along the pencil lines, I glued everything together again so that I could reopen the assembly in the plane of the third dimension, and so see exactly how these interdigitations meandered.

Quite apart from the scientific interest of what we found, I was gratified to see that some examples were more sexually evocative than my original diagram. Furthermore, the receptive partners had occasionally bitten off some of the invasive processes; this had interesting implications for interneuronal information exchange. Stick that in your pipe and smoke it, thought I, as incontrovertible facts went off to press (Boyne and McCleod, 1979).

Wanted: A Real Job

After my first year in Iowa, my one-year contract was renewed. I was still an Associate Igor, now nearing five years of postdoctoral wandering and still not on the tenure track. Yet I was publishing good research in the best journals and felt confident, so I set about finding a real position.

With no famous sponsor willing to recommend me, I was a scientific orphan. I had to depend on ten-minute presentations at conferences in the hope that someone from a recruitment committee would be in the audience, someone who would decide that I was an appropriate candidate for an assistant professorship.

I put together a proposal for synaptic vesicle discharge that was based on the effects we had seen in calcium-soaked nerve terminals. My images supported the possibility that vesicles might open-and-close-and-refill many times before a final collapse. This was a fairly compelling counter-argument to the Heuser-Reese model, which was called "classical excocytosis" and which was quickly becoming the accepted perspective. Arguing with the establishment is a good way of getting attention. Turning that attention into a job offer, however, was still a challenge. In another clear recollection, I can see myself hunkered down in the corner of a bar, the night before I would speak, taking stock while nursing a single beer.

Although the Liverpudlian Irish disrespect all forms of authority, I had still been discomfited by the prosody with which my fellow countrymen can project disdain. Former Prime Minister Margaret Thatcher's tone of superiority is a good example. To my ears, it is like fingernails scraping a slate blackboard. After a decade in America, in a culture with a similar language but a different set of intonations, I had eventually realized that my brain just didn't recognize the local status static. Now, however, I was about to present a public challenge to a prevailing hypothesis, and I was just a postdoc with no faculty position. Drawn swords were to be expected.

As I viewed the tactical problem, I felt that the name of the classical proposal—exocytotic collapse—was a disadvantage. That Latinate term provided an aura of respectability that obscured inconsistencies in the data. I decided that I should first rename the established model. I recalled hearing it described as the "throwing up hypothesis for synaptic vesicle discharge." That seemed to have the Anglo-Saxon resonance that I wanted. Variants were readily achievable—the puking model, etc. Okay, that would work. Next, I had to give a name to my alternative, one that would elevate it and me. This was hard; I couldn't match exocytosis.

My notion was that the ATP of the vesicles might survive the early rounds of acetylcholine release, and foster refilling for an unknown number of times before final collapse, at which point all the contents would be discharged simultaneously. I decided to call it the 'selective secretion' model of vesicle discharge.

The existential gauntlet of a question-and-answer session is an essential part of being a scientist. Surprisingly, perhaps, I did not feel that I needed any luck to get me through it. As I tried to explain earlier, I wasn't claiming to be 'right'; my data provided no definite proof of anything. But the speculation I was drawing was reasonable and it differed from prevailing assumptions. Furthermore, the rough experiences I had already survived gave me the confidence that I was just as intellectually dangerous as anyone else. And this time, I wouldn't crap in my pants.

The next day, when I mounted to the podium, I immediately recognized Tom Reese in the middle of the room; the game was afoot. With every word polished and with innumerable nuances foreseen, I clicked through the micrographs and laid out my alternate ideas in exactly ten minutes, then turned to the moderator. She spoke into the microphone. "Does anyone have any questions for Dr. Boyne about this interesting hypothesis?" I could have kissed her for that editorial comment, but my attention was back on Tom. I waited.

Nobody else raised their hand or moved to a microphone. I realized that the thought line was so clear that there really wasn't much to ask about. Tom seemed uncomfortable. He looked around, plainly hoping that someone else would take up the cause, giving him time to think. When no one did, he rose from his chair and walked to a microphone in the aisle. "Tell me, Dr. Boyne, how do you reconcile your argument with the evidence that one synaptic vesicle disappears whenever one quantum of neurotransmitter is released?"

Even with my low sensitivity to American nuance, I perceived the

Ivy League superiority with which Tom decorated his grammar, his instinctive attempt to put me on the defensive. Yet he was referring to his own and Heuser's work, (Heuser and Reese, 1973), and they had only claimed 'an order of magnitude correlation' between the number of vesicles disappearing and the number of quanta released.

That meant nine selective secretions for every puking event would be completely consistent with their evidence. In other words, my proposal could be the dominant form; it was Tom who should have been on the defensive.

At that point, I don't think even Margaret Thatcher could have gotten the better of me. I took my time, leaned in to the microphone and spoke with a tone as archly English as a Liverpudlian can muster. I first made it clear that I had recognized Tom and knew to whom I was speaking. Then I flatly contradicted the claim that he had proven the loss of one vesicle for every secretion event, and cited several other contradictory results.

In a follow-up question, his tone had changed; I had been granted standing. I responded with a similar sense of "we're all in this together" good fellowship. As I left the hall, I found myself on the receiving end of an invitation to negotiate for an Assistant Professorship in Mississippi, and another in North Carolina. My ship was coming in.

In the end, that wasn't how it happened. In a letter to *Nature*, Silinsky and Hubbard, (1973), had reported that nerve endings in the mouse diaphragm, known to squirt acetylcholine, simultaneously released ATP. This fitted well with our Cambridge discovery. I wrote to Eugene Silinsky, for it was his thesis work—and congratulated him. The letter reached him in Northwestern Medical School's Department of Pharmacology, in Chicago, where he was a new assistant professor. He replied to me in Iowa, with an invitation to visit.

Soon thereafter, a girlfriend was planning to go back to Chicago to see her family, and we arranged to meet Dr. Silinsky.

Gene is irrepressible and unpretentious; we communicated well and enjoyed each other. Towards the end of our visit, and in an unexpected surprise, he asked if I would be interested in teaching central nervous system pharmacology. His department was looking for someone to specialize in that part of the second year course for medical students. Of course I was interested.

A few weeks later, I came back to Chicago to meet the rest of the faculty and to give a seminar. Les Webster, the department chairman, met me for breakfast in a Chicago faculty club, which left me with a mahogany-paneled memory that is somehow convoluted with John O'Hara novels. Les was both a fellow biochemist and an M.D. His directness made communication easy.

With astonishing speed, he seemed to make up his own mind that I matched the requirements, then he shepherded me through the process of general acceptance. As I look back on it, those were all fairy tale events. Suddenly, after five years of samurai wandering, I was a retainer in an academic castle of some standing.

Chapter 16
Behavior of the Lower Organisms

The anthropomorphic tendency imagines human thoughts in other animals and even inanimate objects. Tolerated in children, it is frowned upon in adulthood.

A contrasting impulse insists that it is beneath human dignity to make serious comparisons between human behavior and that of the lower animals. This is hubris, which should also be frowned upon in adulthood. (I said that.)

A Victorian Classic

Les Webster, our chairman, invited Gene Silinski and I to teach a graduate course in behavioral pharmacology. Gene had the neurophysiological side of the issue well within his grasp. I was less familiar with the body of pharmacological knowledge.

Since we actually don't know how normal behavior is produced, I decided that the pharmacological sub-discipline remained on tentative ground, and felt free to forge my own approach.

I remembered a book picked up at an overstock sale in Cambridge, a reprint of a classic text by an American scientist named Herbert Spencer Jennings: *Behavior of the Lower Organisms*, (1911). It had struck me that if single-celled organisms had behaviors, they were surely setting a baseline against which any advantages of neurons and brains in higher animals might be evaluated.

When I finally took the book from the shelf, still unread, I found that my instinct was seventy-five years behind the times.

In the foreword to the 1962 edition, the psychologist Donald Jensen reviewed the intellectual turmoil that Jensen had generated with his rejection of his society's preferred viewpoint.

I felt as if I had found a ghostly mentor, for when microscopic animals had been discovered, Victorians were astonished to learn of their hurrying, scurrying ways. But many of the people who then rejected the idea of evolution were looking for something to be scornful about. The apparent notion that 'infusoria' might display consciousness and free will was a perfect target.

Moreover, theologians had long insisted that amongst animal kind, man alone had been provided with a soul—and that this was the source of free will. The implied conclusion was the nice one-liner that other animals were 'just' bundles of reflexes. Those who accepted evolution, however, were actually demanding answers to a more subtle question. *Did animals establish precedents for human behavior?*

Mary Shelley's *Frankenstein, or, The Modern Prometheus* (1818) had spawned a genre of mad scientist fears, and Jennings mentor in science, Jacques Loeb, may have feared stirring popular anger by making any demeaning comparisons between man and animals.

Scenes of local villagers bursting into laboratories are a natural nightmare for scientists. Whatever his original motive, Jacques Loeb strenuously advocated that human-associated terms be avoided when referring to animal behavior.

Once Jennings began his own research program, he developed an independent view: that Loeb's purported objectivity was just the opposite: a bias. I will take the same position, and so I need to be clear in distinguishing the anthropogenic impulse from what Jennings pioneered. The Australian snake-necked turtle is a splendid oddity of a beast. She has horny jaws fixed in a permanent smile, which is quite endearing. It is obviously misleading, however, to associate that rigidly curved jaw line with any particular emotion: that would be the anthropogenic mistake.

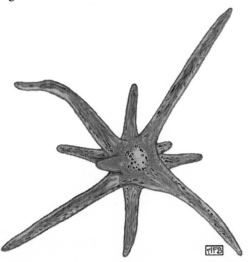

Now consider that when the shape-shifting creature called amoeba is suspended in water—with no surface to move upon—it uses virtually all of its protoplasm to generate an array of pseudopodia that extend in random directions, as shown above. In stating that the amoeba "uses" all of its protoplasm in this way, I have already violated Loeb's sense of objectivity: "use" is usually associated with a thoughtful process.

The difficulty compounds: when one of the pseudopodia touches a surface, the others retract and the whole animal flows into the pseudopodium that has "successfully located" a surface.

Although Loeb was adamant that such terminology violated objectivity, Jennings chose to ignore him and to use well-known human words for apparently similar animal behavior. The immediate academic response was infuriation, particularly from Loeb and John Watson. This same Watson would become a founder of behavioral psychology, the discipline that B. F. Skinner carried forward, and which tried to squash all considerations of subjective experience.

Behaviorism has now been eclipsed by cognitive psychology, which acknowledges the phenomenon of consciousness.

Despite the furor, the clarity of Jennings' descriptions and the evident objectivity of his experiments provided scientific high ground. When the dust finally settled, Jennings was still standing and Watson had conceded that human behavior is rooted in the evolutionary past. Loeb appears to have carried his disagreement to the grave.

In attempting to keep animal and human behavior separate, pride and status sensitivity had limited Loeb's intellectual achievements. These problematic human characteristics appear to have no counterpart in the microscopic world. They will concern us again.

An Ancient Principle: Trial and Error

Jennings began by showing that single-cell cytoplasm responds to all the stimuli that affect man himself—changes in pressure, temperature, light, gravity, acids, salts, and electricity. (Jacques Loeb would later show that even non-living colloidal material responds to these forces.)

Jennings went on to show that the so-called reflexes of simple animals were more varied than previously believed. He defined their fundamental behavioral principle: *Selection of the (favorable) results from varied behaviors.* By "varied behaviors" he meant trial and error.

Two sets of Jennings' observations will help put animal consciousness and the trial-and-error principle into a broad contextual frame.

The Success Rate of Trial and Error

I drew the floating amoeba with ten pseudopodia. In such a case, when one reaches a surface, the other nine will retract. That amounts to nine errors made in pursuit of one successful outcome.

Exercise of the trial and error principle typically produces such a preponderance of failed attempts. This dominance of error in trial-and-error tactics may seem a trivial point, and it did not merit emphasis in Jennings' work. Nevertheless, it will become central to our understanding of how our cerebral hemispheres constitute a rationality engine. But for this chapter, there is more to learn from 1905.

Jennings made a still more disconcerting discovery: the reflexes of single-celled creatures are deployed in a flexible order. To appreciate why this was a philosophical blasted nuisance (to Jennings' peers) we need to join the master at the microscope and consider some specific examples.

Stentor

Stentor is a millimeters long, single-celled protozoan with a mucus-covered foot that is generally fastened to an aquatic plant or a rock. Its trumpet-shaped body extends into the surrounding pond water. Around the wide end of the trumpet, multiple cilia are fused into paddles that sweep food into its gullet.

One of the earliest efficiencies in behavior is pattern analysis of irrelevant noise in the environment, in other words, sensations that signify neither harm nor benefit. Compulsive response to these is logically a waste of energy, and Jennings tested whether Stentor had evolved such logic. He brought up the tip of an ordinary glass pipette and blew a puff of harmless water at an individual Stentor. The creature bowed away from the jet, repeating the effort for three or four puffs in succession. Then it quit responding.

The behavior is called habituation. It is a natural wisdom that can be built into a single cell; it does not require neurons and a brain. In human parlance, an ability to habituate would be called 'an ability to ignore distractions.'

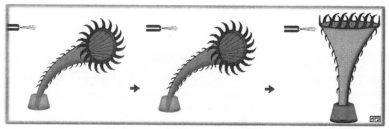
Stentor habituates to a water jet

Although we inevitably involve our brains and some fancy circuitry when we concentrate, we are not entitled to claim that 'habituation to irrelevant stimuli' is special to human mentation. Whatever icing we have added to our version of the cake, the basic recipe is ancient.

Next, Jennings tested whether Stentor would habituate to a noxious stimulus. He put acidic carmine grains in his pipette and blew them at the previously habituated individual, which immediately restored its bending-away behavior. Jennings waited for it to return upright and blew another puff. In this circumstance, it did not habituate; it kept bending away. After several more puffs, the creature deployed another tactic: it bent away and simultaneously reversed the beat of the mouth paddles—and so actively drove away the carmine! This is what gave theologians the willies.

After several further cilia reversals with avoidance bends, continued puffs led Stentor to scrunch down inside its mucus cave. But each time it reemerged, Jennings delivered more carmine, to which it responded by staying longer and longer in its cave.

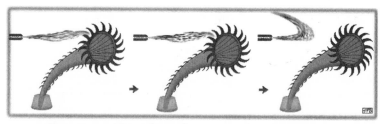
Stentor avoids and repels carmine-laced water jets

Yet still the valiant Jennings kept on puffing—until another response was elicited: Stentor dove back into its cave, pulled up its foot from the rock, burrowed out through the side of the mucus wall, puckered up its oral cavity, and swam off, now using its body cilia for locomotion, evidently 'concluding' that the old neighborhood had gone to hell and 'searching' for a new place to settle.

Although I feel confident that Stentor was not thinking anything, while it was choosing to hide and then escape, I also believe that when I was thinking of escaping from a wiper's position in the engine room of a steamship, my impulse had evolved from the same properties of living protoplasm.

The Origin's of Self-Awareness

Jennings' studies ultimately led him to challenge earlier attempts to define consciousness. The problem was that, even for animals as simple as Stentor, the various options for avoiding carmine grains were not deployed in a rigid, fixed sequence, as one might expect for a mindless series of reflexes.

Instead, on different occasions, the response sequence was different. "Selection" from a behavioral pallet of different options, however, fulfilled the Victorian definition of consciousness. Jennings, with great diplomacy, suggested that we revisit that definition (presumably to stay ahead of the riff raff).

He also spent many hours observing the hunting behavior of amoeba and concluded that if we met a man-sized amoeba in its hungry mode, its demeanor would connote as much consciousness as we readily grant to a wolf. One might add that barring a door with a gap at the bottom would offer no protection against amoeba attack.

These were the sorts of thoughts against which Jacques Loeb had attempted prior ego-heavy restraint. Yet they are appropriate and rational thoughts; when that is appreciated, they lead to substantial new insights.

Damasio and Self-Awareness

It has been very difficult to explain our own awareness of being conscious. This is because our explanation process typically relies upon a combination of direct experiences and metaphoric comparisons. But our own self-consciousness is a uniquely direct experience; anyone else's claim of self consciousness can only be inferred. Furthermore, mutually agreeable metaphors have been hard to formulate.

The notion that there is another little man inside our brains who looks at the data from the eyeballs is a rather pathetic attempt to use the lone instance we have: it leads to an infinite regress of increasingly tiny men—a self-extinguishing explanation.

Nearly a century later, a completely different approach to comprehending consciousness has left the station. Daniel Dennett (*Consciousness Explained*, 1991) emphasizes that the neuroanatomy of the brain produces innumerable opportunities for electrical signal traffic to travel around in loops, gradually accreting additional insight. He considers that such recursive anatomy allows recursive thoughts to place details into progressively larger frameworks of broader contexts. I like the metaphor of a mental onion that sprouts new layers enclosing all that went before, and so providing a broader perspective. Dennett suggests that the brain found survival value in constantly reporting to itself (between the layers of the onion), and that this eventually produces the consciousness of which we are conscious.

Douglas Hofstadter, in *I am a Strange Loop* (2007), has suggested that with enough complexity, any information system should reach a threshold at which it is able to perform a Mobius Twist, and examine itself, i.e., become conscious. I favor a natural extension of Jennings' perspective, which can be found in Antonio Damasio's, *The Feeling of What Happens* (1998). Damasio uses simpler creatures as his metaphorical starting point, and he emphasizes the neuron's first role in evolution: to sense the state of the body—the proto-self—and organize reflexive responses to sustain the internal stability upon which life depends.

Additional rounds of higher-order-sensing inevitably introduce feedback information on the effectiveness of the responses to the first round of feedback. (An example: our brain senses when our muscles need more oxygen, and responds by accelerating both respiration and the heartbeat, which delivers the necessary oxygen; the summary report on this initial effort is whether we still 'feel' out of breath.) After millions of years of evolution, enough rounds of feedback reporting may yield a 'feeling' of awareness of being an entity that is aware of its own reports.

Damasio suggests that the last round in the modern human includes knowledge that our memory bank of self history is well worth consulting as we consider what to do next.

The beautiful and appealing part of these ideas is that they convert the effort to understand self-awareness from an infinite regress of little men to an indefinite progress of broader and broader perspectives. At least progress is something we have seen before; we can relate to it metaphorically.

Paramecium

Jennings also reported on the behavior of the microscopic protozoan paramecium. Using synchronized cilia, beating like tiny oars, paramecium swims forward, rotating through a spiral path as it goes. Its ability to negotiate a complex terrain filled with mechanical obstacles is one of Jennings examples of the survival value of trial and error.

We now understand the mechanism at an electrical level. Since paramecium lacks an advance warning system, it will eventually bump into some solid object. At this juncture, mechanoreceptors (i.e., proteins embedded in the lipid of the cell membrane) report the collision by opening a pore, allowing an inrush of calcium ions that collapses the normal cell voltage. The effect is roughly comparable to Mr. Rogers sticking his finger in an electrical outlet, at least in the sense that it affects the protoplasm and changes behavior. In paramecium, elevated internal calcium levels reverse the ciliary motor, which causes paramecium to back up. It looks like a wise move, but it is a reflex.

As the ciliary beat is reset, the new 'forward' direction becomes unpredictable—and this is what gives paramecium the properties of a trial-and-error machine.

Because each forward surge attempts a new direction, a paramecium will inefficiently but eventually 'find' an open path to the next wall, where it will repeat the arduous 'discovery' of a way forward—until it has traversed an entire maze.

Paramecia exhibit another sophistication: their cilia beat in a manner that samples the bulk water ahead of their approach. Any changes in conditions are therefore perceived by the oral groove before the entire body is immersed. If adverse changes are detected, the creature performs the aforementioned retreat, and it then samples the water in a circle around the former direction before 'deciding' in which direction to travel.

Epilogue

There is a well-accepted perspective that we humans use our brains to perform internal trials, selecting the most promising for execution in reality. This is usually noted as a tremendous gain in efficiency and safety over single cell critters, for they must generally discover their mistakes after the fact. While in no way arguing with that suggestion—after all, we all know that we review our options—I want to again draw attention to a different, unflattering corollary.

Since trial and error produces a preponderance of errors, the brain must contemplate many errors, or not think at all.

This inevitability leads to a question:

Have too many mental errors escaped and been implemented in the real world?

The evolutionary scale of reference behind this question positions us just where we need to be: outside the usual box of political partisanship, or knee-jerk patriotism. It implies a species problem on a global scale, with possibly cosmic consequences. I'll try to carry the weight of those assertions in the following chapters.

Chapter 17
God Voices

Julian Jaynes and Schizophrenia

In my effort to perform as a pharmacologist at Northwestern, I felt most insecure about lecturing on schizophrenia. Although I could readily collect all the details from the textbook, I still hated the idea of describing how to treat something with which I had no personal familiarity. I tried to punt by asking the Chairman of the Psychiatry Department to do it for me, and he agreed. When the time came, however, a psychiatry resident turned up, apparently intending to add a teaching claim to his resume.

I introduced our guest and sat down to listen. He began to read from the book chapter! The class grew progressively more restless, and finally he announced to them that they were themselves displaying one of the primary symptoms of schizophrenia: one of the four 'A's—lack of attention. At least he had finally interacted. Of course, I realized that I couldn't be worse than that. In the following year, I would have to do it myself, come hell or high water.

Fortunately, Julian Jaynes' book, *The Origin of Consciousness in the Breakdown of the Bicameral Mind* (1976) appeared that summer.

God Voices

Jaynes argued that descriptions of the presence of God voices in the Hebrew Bible and in the early Greek Iliad should be taken seriously—as evidence that auditory hallucination once played a socially important role. To explain those ancient perceptions, he suggested that evolution of language allowed an impression of speech to be driven from the silent right hemisphere.

Jaynes recognized that the left hemisphere learns to externalize its linear 'thoughts' in the form of sentence sounds. He suggested that, in some individuals, the more broadly perceptive right hemisphere, wielding an internal impression of speech, might have been able to practice internal ventriloquy, thereby 'driving' the self as a subordinate secretary-executive, subject to authoritarian command. He interpreted the archaeological record as suggesting that our inner dialogue had once taken this imperious form.

The notion might be dismissed as a science fiction speculation, and there were many willing to do so. But ancient texts explicitly report such experiences. Moreover, modern schizophrenics still report hearing voices that are more authoritative than their 'selves' or their care-providers. As in the Greek Iliad, and the Bible, schizophrenic voices may even command mortal assault!

Furthermore, recent brain scanning confirms that the auditory hallucinations of schizophrenics activate brain areas that more normally respond to external voices. Since these areas are not active when we simply imagine conversations, the notion that the brain really can generate the bicameral sensation of disembodied voices is no longer speculation.

Remember the question of which half-wit should be steering Lurline's lifeboat? Jaynes was suggesting that the reports of detailed authoritative commands arose when the right hemisphere recognized that it was better tuned to survival imperatives and so issued appropriate commands to the left.

How might the right side have come to a conclusion so offensive to the ego of modern man? Willingness to contemplate many simultaneous variables underlies the right hemisphere's skills in pattern analysis. In contrast, the left hemisphere prefers to follow small scale details in linear, 40 bits-per-second cause-and-effect sequences that it then puffs up with the term 'logical analysis.'

Jaynes suggested that when early verbal humans experienced centuries of stable conditions during which their major preoccupation was with finding food, their accumulated pattern knowledge enabled accurate anticipation of appropriate executive actions. Consequently, right hemispheres expressed authoritarian commands—often associated with rational farming practices. These had clear long-term survival value.

An interesting form of 'hard' evidence is available. Archaeologists long ago found that the fields of Mesopotamia featured stone stele upon which were carved descriptions of how God voices had commanded the cycles of farming in ancient times. Similarly, the moral code of King Hammurabi is recorded on a stele, and it also describes moral pattern wisdom, in this case the code Hammurabi received while listening to his God Marduk: *an eye for an eye.*

Jaynes suggested that the stele themselves were made when the right had commanded the left to reify these voices in durable written form—another example of a long term perspective emanating from the right hemisphere.

Advent of the Verbal Ego

But why and how did the voices fall silent? As times changed, the Greek Iliad seems to record a transition from an impression of 'God control' to the discovery of personal responsibility—self-control. What had now happened? Jaynes suggested answer was in two parts.

The development of writing introduced the personal pronouns *I, me, and mine.* Children who absorbed such word-ideas might have learned to inhabit them more deeply, generation by generation, eventually completing a transition to a personal sense of executive authority, i.e., modern self-consciousness. The biblical suggestion in Genesis—that we learned to know ourselves—is a literally equivalent explanation. Where the Bible, however, posits a metaphorical serpent in the Garden as the trigger, Jaynes had another literal suggestion.

Life in any community eventually meets with a catastrophic event, such as volcanic eruption, earthquake, fire, or tidal wave. One-off events are too novel for pattern analysis by short-lived creatures. During an acute crisis, (Jaynes' candidate was the eruption of Mount Thera with subsequent tsunamis throughout the Mediterranean Sea). Biblical right hemispheres might then have experienced a 'cannot-compute mode'—and so the God voice fell silent. In cultures already using personal pronouns, the modernizing self might then have found that its verbal strings of short-term logic were well adapted to day-to-day survival. We can view ourselves as still being in this phase. As we go about our daily routines, however, mostly surviving from paycheck to paycheck, a long term negative consequence has been building up: global pollution and planetary degradation.

Our left hemispheres seem insensitive to the associated long-term considerations. What about the right? For most of us in the twenty-first century, the whispering voice of conscience—a somewhat tattered remnant of the ancient God voices—is all that reaches the clear consciousness of our subjective selves in the left hemisphere.

The advent of prayers, part of the historical record noted by Jaynes, in which God is beseeched to return and manage our affairs once more, suggest that the transition to deepened personal responsibility was accompanied by some regret.

Although these were magnificently original thoughts, Jaynes had difficulty finding a publisher, and the reasons are interesting.

Professional response

Academic presses sent the manuscript to better-known psychologists—who reviewed Jaynes' thesis with reductionist derision, typically asserting that the notions were beyond proof. (How could you 'prove' the subjective experience of an ancient Greek? You can't. But does that mean one shouldn't contemplate and discuss the possibilities?)

I suspect that those reductionists had swallowed behaviorist dogma and so thought that Jaynes' discussion of consciousness was a fool's errand. Another possibility is that they were defending their discipline from an implicit criticism: if Jaynes was right, then they, the supposed experts, had all been overlooking hugely significant psychological assertions in well-known, ancient texts. In response to such professional skepticism, the publishers turned down the manuscript. (Others using brain research to answer basic questions about our behavior have also met resistance from entrenched scholars. Candace Pert gives a compelling account of many examples in her book *Molecules of Emotion—The Science Behind Mind-Body Medicine* (1997).)

Like many others, I first learned of Jaynes' message in a *Time* magazine review. A nonacademic press, Houghton Mifflin, had initially lost Jaynes submitted ms, but when they rediscovered it, they agreed to publish in 1976. Practicing due diligence, the *Time* reviewer had again solicited professional comment, and it was again negative, but it was so unprofessionally catty that the reviewer cited the remarks directly. One psychologist simply asked, "Who is Julian Jaynes?" This was a slur on the fact that Jaynes had published very little in the way of reductionist science. Another was more blunt, declaring that Jaynes was obviously clever, but expressed a wish that he had done something "more useful with that cleverness."

Recognizing the greenest form of professional envy, and needing to develop my lecture on schizophrenia, I bought a copy of Jaynes' book right away, despite what I thought was the most horrendous title I had ever seen. So did many other people, and it has now been continuously in print for thirty years, noticed and read by the general reader. I can't remember the names of the critics.

Chapter 18

The Story of an Invention: Gentleman Jim

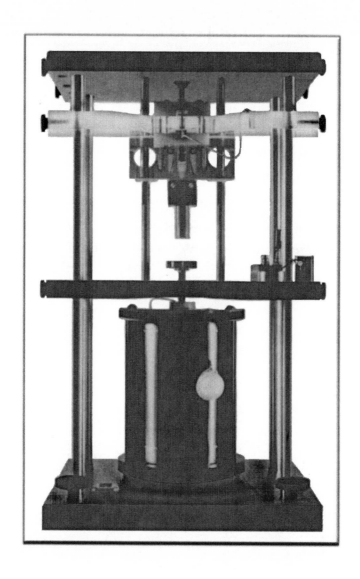

Quick Freezing

As an Assistant Professor, I was expected to bring in my own research funds. To facilitate the effort, Les Webster had arranged some seed money with which I could get started, and the Dental School at Northwestern had bought me an electron microscope, housed in one corner of my substantial laboratory. Encouraged by Les, who referred to Gene and I as his future superstars (better than some of the other names I'd been called), I wrote a grant proposal to the National Institutes of Health (NIH). I proposed to focus on the manner in which synaptic vesicles discharge through nerve endings in the electric organ. Though invisible to the naked eye, this is a holy grail phenomenon because it seems likely to be part of the adjustable-synapse biology that underlies all learning, memory, and thinking. Shortly after funding was awarded, however, none other than John Heuser and Tom Reese rendered the technological basis of my proposed research obsolete. I'll first explain that.

I had traveled to a meeting in St. Louis, and learned that Heuser and Reese had solidified synaptic vesicles in the very act of discharging, using a technique called 'quick freezing.'

The idea had arisen in the 1960s from a man called Van Harreveld, who was a laboratory neighbor to Roger Sperry at Caltech. Van Harreveld had reasoned that sufficiently fast cooling ought to prevent the water molecules of living tissue from aligning in crystalline ice formations. Theoretically, they would have no choice but to form a solid liquid—a glass. He thought that this would preserve the architectural details of the living state perfectly, and permit high magnification electron microscopy. Unfortunately, he calculated (correctly) that the required rate of cooling was 20,000 degrees centigrade per second.

I once made the mistake of thinking that a warm-blooded animal dropped into liquid nitrogen would die instantly. It turns out that a warm animal radiates heat so powerfully that liquid nitrogen cannot make contact before the liquid is vaporized. Consequently, a relatively insulating, counterproductive cloud of gas instantly surrounds the animal. This gives plenty of time for the tissue water to crystallize, and so disrupt cell structure. (It thereby slows the dying process, so it is not an ethical way to kill hairy mammals.)

Although he could not afford to lose freezing power to a cloud of gas, Van Harreveld had one advantage: he only wanted to freeze tiny samples that would later be magnified many thousand times in an electron microscope. He calculated that the necessary rate of cooling would be just theoretically possible if he first chilled a silver bar in liquid nitrogen while holding its end, like the top of an iceberg, out of the liquid.

Once the bar was at the same cold temperature as the gas, a small piece of living tissue could be touched to the dry metal surface, and freezing should occur super fast—both because the metal wouldn't vaporize and because it has a high thermal conductivity. He made a simple apparatus that included a spring-loaded plunger stick to make sure that the tissue made rapid contact with the silver iceberg. (A slow approach might allow those dastardly ice crystals to form before impact.)

I have always felt deeply grateful to Van Harreveld for describing his laboratory experience accurately. He reported that nine times out of ten, the attempt failed, and later analysis showed that the tissue was full of crystals—useless. The tenth time, of course, was the kicker, and he published beautiful examples showing that the natural state of brain tissue was much less swollen and more orderly than it ever appears after using the standard aldehyde embalming techniques.

In fact, when quick-freezing was successful, nerve endings finally looked as though they might support intelligent life! (Van Harrveld and Crowell, 1964.)

Now we can return to Heuser and Reese at NIH, who decided to apply the technique to the frog neuromuscular junction and the question of synaptic vesicle behavior. They attempted to increase the frequency of successful freezing by changing the cooling fluid to liquid helium, which vaporizes at just four degrees above absolute zero, the coldest temperature possible in this cosmos. They also changed the delivery system so that electromagnets turned on and clamped the sample in contact with the metal surface. The colliding magnets generated substantial noise, and the instrument acquired the nickname *The Slammer*.

Their earlier research was widely interpreted to mean that synaptic vesicles open and evert into the terminal membrane, disappearing in the process. This idea is easy to illustrate, as shown on the next page, and the textbooks had adopted the proposal with gusto. But it was all based on using slowly embalmed tissues, and thus questionable.

If one used a broader perspective, there were many indications that synaptic vesicles might have other modes of discharge. I suspected that overstimulation caused the vesicle-destructive version shown overleaf, and that a more usual physiology was to simply open and close in place, allowing the vesicles to refill and secrete again.

Heuser and Reese attempted to capture the above sequence in incontestable detail using the new technology, but for a long time, they did not find what they wanted. They might have concluded that vesicles were indeed using a less visible method of secretion, but that would have conflicted with their earlier work. Instead, they did a calculation: it suggested that the efficiency of neurotransmission required so few vesicles to open that they were just hard to find.

This line of reasoning opened the way to adding drugs to force a hyper-secretive state on the nerve endings. Once they increased the rate of secretion by about 200 times, they found the-long anticipated pictures of opened vesicles in the discharge sites, and they looked beautiful! (Heuser et al., 1979.)

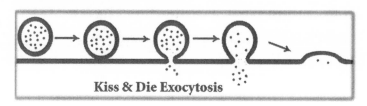

Kiss & Die Exocytosis

This time it was my turn to sit and watch John Heuser present conflicting data. I was aware that his interpretation did not exclude the opening and closing process, now called kiss and run (and refill). For those of us directly experimenting on the issue, the notion that the drug and salt conditions they had chosen were forcing the terminals to behave unnaturally was perfectly cogent. However, the technology and the pictures were so superb nobody was ready to hear criticism.

The images encouraged the wider community to close ranks concerning this long-running question, and accept the textbook presumption. This time the dominant spin was directed at 'nihilists'— meaning people like me. I decided that retreat was the better part of wisdom, and asked no questions. I took my doubts back to Chicago.

Bounce Prevention

My assets were a three-year appointment, a lab, an electron microscope, and three years of funding from the NIH. The liability was that it was no longer worth doing any of the proposed experiments (because Heuser and Reese's work left my planned formaldehyde fixations looking inadequate to the task, a viewpoint that I immediately accepted).

One doesn't want to give the gummint the grant money back in such a situation. I thought about whether to call John or Tom and ask for blueprints of *The Slammer*, so that I could copy it, and continue trying to prove that their experiments were misleading. I thought the conversation would feel awkward, and decided against it.

Another instinct told me that they had used liquid helium out of a quaintly Victorian attitude: *If I go to a lot more trouble, surely I will be rewarded with more success.* Liquid helium is expensive and has curious abilities to flow upward against gravity, which forces use of expensive containment vessels. My grant had not budgeted for any of that. Furthermore, liquid nitrogen had worked for Van Harreveld, so cheap techniques had already been proven. It was just a question of getting them to work reliably.

Two other workers, Joe Pysh and Ron Wiley had also demonstrated shape changes in stimulated nerve terminals; they had studied aldehyde-fixed cat spinal cord nerve terminals. Furthermore, Joe, who was in the Anatomy Department at Northwestern, had acquired a replica of the Van Harreveld's freezing machine, which he graciously lent me. Compared to Heuser and Reese's Slammer it looked like the anemic kid on the beach, the fellow who can't get a girlfriend. I played with it at room temperature, not bothering with cryogenic liquids, just watching the delivery system operate. It wasn't smooth, needed plenty of lubricant, and the adjective "rickety" rattled to mind.

Deciding to monitor the impact event, I mounted a blunt pencil on the end of the delivery probe, and connected the central carbon core to an electrical circuit. I wanted to "see" the impact event as an electrical phenomenon, so that I could later extend the data in time and study it carefully. To do that, I borrowed an oscilloscope from Gene Silinsky. As soon as I watched the pencil fall, and turned to the oscilloscope screen, I knew what the problem was.

The observation was simply this: the circuit had closed and reopened multiple times in the first few milliseconds. This meant that the pencil was bouncing at impact, before the two-thousandths of a second that was required for frozen-glass formation, and too fast for the eye to see or the ear to hear. The tissue needed to be delivered in a dampened manner.

Non-mechanically inclined readers will not be able to follow the specifics of the following description, but it is only the pattern of thinking and rethinking that I am trying to make vicariously familiar, and you will perceive that much.

Virtually all solid objects bounce apart when they first collide: crockery on a countertop; clutch plates in a car; closing doors, etc. In none of these cases does the first bounce, in the first millisecond, have any critical significance. It was the attempt to freeze living tissue that introduced a new level of rigor. The goal of preventing that first bounce was an appealing problem because it might allow me to make a mechanical gizmo without needing a background in engineering. Furthermore, application of quick freezing to electric organ might yield evidence for the form of synaptic vesicle behavior—open-and-close—that had been the original point of my grant proposal. And I could use that possibility to justify not giving back the grant money. This overall situation seemed to justify every late night with the books, and every hoop I had jumped through for two long decades. (It might seem that my efforts at this point had little to do with bullshit, but remember that all thinking depends upon the behavior of synapses—including the bullshit.)

I went home early, prepared supper, and sat in an armchair, twenty-seven floors above the shores of Lake Michigan, stroking my cat. We often sat together and watched the dark storm clouds that rolled out of Canada. As they crossed the lake, they seemed to rear up and hurl lightning bolts at the John Hancock building, the Sears'

226

Tower, and the Playboy Club. Pongo would have loved it.

There were no such distractions this night, and I began to consider delivery options. Eventually the notion came that a two-part system might be arranged so that the impact of the first part was smoothly coupled to continued descent of a heavier second part, which might overwhelm the recoil of its predecessor.

My first mental design included a brass plunger traveling in a barrel full of hydraulic fluid. I realized that, upon impact, the plunger, continuing to descend, would squirt fluid every which way. This would look really hokey and make a mess in the lab, but it seemed important to ignore those complications. In other words, it seemed important to leave my status sonar switched off, perhaps so that my forty-bits-per-second bandwidth of consciousness was free to concentrate on whether the general principle was valid and practical: *Would the first bounce be prevented?*

Growing confident in that regard, I suddenly realized that I could build in an air pocket to collect the squirting liquid before it made a mess (and thereby raised doubts about the inventor—at this point I had let status concern into the picture).

I now saw something unexpectedly sophisticated: with a sealed air chamber deliberately in place, the energy that formerly generated bounce would now go into compression of that enclosed air bubble. Furthermore, after reaching maximum compression, the air would re-expand, forcing the hydraulic liquid and the brass plunger back where they had come from, while prolonging the downward pressure for more milliseconds and simultaneously resetting the whole thing for another freezing!

The point of this description was to show the flow of mental trial and error in the creative process. Each element is simpler than legend commonly tells; the effort only becomes laborious because one must keep going back and forth so many times.

We seem to vary in the number of such iterations up with which we are willing to put. I was glad that the mental sausage making would not be obvious in the final technology. Instead, it would probably seem quite sophisticated, which is an advantage for a young Assistant Professor. It might even make women think that I was some kind of Albert Ape, and ...well I've probably said enough about that.

Here is a drawing that may or may not be of interest: your choice.

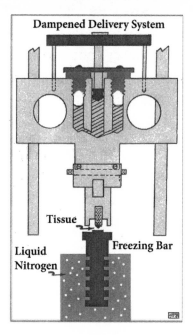

There was another odd component of the experience. Growing confident that I had a stable mental picture of a device that I could make with my limited machinist's skills, I went to bed convinced that it would work. I woke early, ambitious and excited, and walked the lakefront to the medical school, stopping at the biochemistry department to describe the idea to a friend, Dave Schwartz.

Strangely, my first verbal description left me feeling unconvinced. This is called cognitive dissonance. While originally meditating on spatial relationships between moving parts, enthusiastically using my imagination and deliberately suppressing status sonar, I may have been emphasizing the spatially adept silent hemisphere.

The neural modules therein seemed to believe. Nevertheless, as I put the mental video into words, and the self-conscious verbal hemisphere became dominant—I no longer believed its rationalizations!

I retreated to my office and found that my belief systems continued to seesaw. The only option left was real-world trial and error: to build a prototype and see if it performed as desired.

The department had a small machine shop, and I learned to use the lathe, with which I fashioned a Teflon tipped plunger and several other small parts. I needed a stout base and found an overbuilt breadboard in a nearby hardware store; it soon sprouted lengths of galvanized plumber's nipples. The prototype had a Neanderthal quality, and I came to regard it as *Grandfather Jim*. Neanderthal or not, according to the oscilloscope, the final assembly was rock solid: the sample tip didn't bounce when it hit the metal bar! It was time to sacrifice a fish.

In Texas, our supplier, Walter, scoured the Galveston Bay area for electric rays and then shipped them to us by air in plastic bags filled with seawater and inflated with oxygen. Although they traveled in a Styrofoam chest, it was winter, and when I met them at O'Hare airport, they were floating upside down, and so cold I thought them dead. I put my hand in and righted one, feeling like Gollum, mourning the loss of his precious.

Then I saw that the dorsal spiracles were still opening and closing, but very slowly. Racing the fish back to the lab, I floated the bag inside the final aquarium for about an hour, while warming took place. The spiracles accelerated, and some voluntary motions returned.

We have been discussing the electrical patterns in the brain that drive thought and movement. The system can be messed with.

Thinking that I could repeat the gentle handling I had done at the airport, I tried to manually ease the warmed-up rays from the plastic bag into the aquarium proper. As the first ray's discharge activated my entire arm, 'I' suddenly became an extension of a fish's brain and so behaved like a puppet manipulated by a puppet master. Biceps proving stronger than triceps, my arm flexed, flinging seawater all over the ceiling.

The fish slipped away and settled to the aquarium gravel, a smug look all over its slimy face.

The evolved 'intelligence' of nerve terminals

We removed many stacks of electric cells from one of the fish, quick-freezing them with *Grandfather Jim* and keeping them in dry ice while cold acetone dissolved the solid water overnight. After that, each stack was embedded in a separate block of plastic, using well-established embedding techniques.

Some time later, my lab assistant, Susan McLeod (now M.D.), was examining the samples. The first block was full of crystals; it would turn out that the collagenous nature of electric organ encouraged a drop of liquid to collect at the freezing face, wasting most of our freezing power. Two steps forward, one step back, a common experience. In the middle of the afternoon, however, Susan suggested that I take a look at another block.

When using an electron microscope, one shuts the door and turns the lights out. Using a pair of binoculars, one peers through a leaded-glass porthole at the shadow cast by a magnified spray of electrons. These are the electrons that have passed through the tissue sample. They cannot be brought out, for they won't penetrate the glass of the view port. However, there is a neat trick that compensates: by having the electrons hit a screen bearing fluorescent paint, a corresponding light image is generated inside the microscope. Since light photons readily pass out through the glass port, one can see what's going on in there.

As I peered down, I saw synaptic vesicles aligned in rows with the membrane through which they were supposed to secrete; these were evident "active zones." This had not been seen (in this tissue) in twenty-five years of aldehyde-based electron microscopy. One active zone even included a smaller than normal dark vesicle that I suspected might have just opened, taken on a load of extracellular calcium, and then closed.

In those pre-digital days, one exposed a negative inside the electron microscope by lifting the fluorescent screen (by way of a set of levers), allowing the electrons to land on a photographic emulsion instead of the fluorescent paint. In a superstitious attempt to ensure that the electron-generating filament at the top of the microscope would not choose that moment to break, I held my breath as I took each picture. I didn't want to be left with a fish story of images that got away, assertions for which I would have no data.

Once that was done, it was time for a major piece of context analysis. At that moment, I knew a fact of electric organ nerve terminal organization that had high explanatory value: a subpopulation of synaptic vesicles (it would prove to be seventeen per cent) was docked at active zones, ready to secrete. The image also suggested that our techniques could be used to follow the putative opening and closing of synaptic vesicles.

Such moments make up for a great deal of drudgery and trouble-shooting at the bench; they are a scientist's version of the view from Mount Everest.

While vesicular opening and closing is now known to be real and normal, I believe the above image is the first example to be photographed. As such, it probably deserves broader recognition than I have previously been able to generate.

Jung once said that we spend the first half of our lives trying to establish our individuality, and the second half trying to reintegrate ourselves into society. My first half was over. I felt sure that I would gain tenure. I sat in the dark for quite a while, and smelled the roses.

Although my time at Northwestern was idyllic, the previous five years, my wandering samurai postdoctoral period, had amplified the institutional distrust I acquired as a graduate student. Having built a working prototype of a scientifically desirable instrument, I knew I could produce more. It looked like that could be Plan B.

Tom E. Phillips (now at the University of Missouri, Columbia) began his Ph.D. thesis work with me at Northwestern and continued at the University of Maryland's Medical School in Baltimore.

We realized that the acetone-freeze-substituted vesicles sometimes seemed to have a grey internal matrix, which left us with a sense that we were on the edge of holding all their natural contents in place. Trying to go the extra mile, Tom began to place the frozen tissues in a liquid called tetrahydrofuran, one that would dissolve ice but very little else. (This was a suggestion from Dick Ornberg and Tom Reese.) We were rewarded with images of cholinergic synaptic vesicles that were clearly stuffed with something, as shown in the next micrograph.

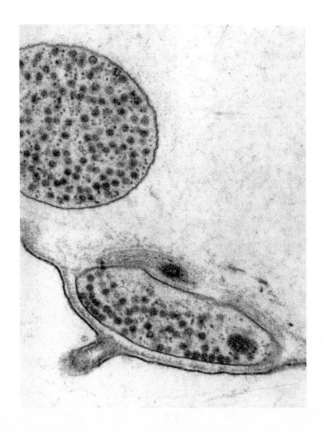

I've mentioned that quick frozen tissue preserves the rationality of nature. The rounder profile at the top of this image is a thread of axon cut in cross section; it is filled with jumbled synaptic vesicles that were produced by the parent neuron cell body, far away in the fish's brain. These synaptic vesicles then ride a system of micro-railway lines towards the nerve endings where it will be their task to discharge meaningfully quantized packets of neurotransmitter. At the level of this round profile, they are not meant to discharge for there is nothing sufficiently nearby to receive the signal.

The lower oval profile, however, shows a nerve ending applied to the face of an electricity-generating cell of the electric organ. In this location, vesicles finally align themselves in squadrons that can efficiently discharge across the synaptic gap.

This level of easy-to-perceive rationality is destroyed in aldehyde-embalmed tissues. In fact, aldehyde embalming selectively eliminates those vesicles docked at the active zones, leaving considerable doubt as to how acetylcholine gets squirted out. Thirty years later, it is widely but still not universally accepted that vesicles secrete acetylcholine in response to an inrush of calcium from the extracellular fluid, and that opening and closing events can be expected in vesicles docked at active zones.

Is it really important to go to all the trouble and expense of learning exactly how neurosecretion happens? I have noted that the synapse is the basic site at which thought takes place; the spot where memories interact with current circumstances, the place from which any potential for resonance will be discovered. Psychoactive drugs seem to have their primary effects here.

If we ever hope to design psychotherapeutic drugs rationally, rather than discover them by controlled accidents, we will need to understand exactly how the system works and which components may be most amenable to drug-mediated adjustment. So yes, a complete understanding of synapses is very important to anyone that ever appealed to a drug for mental assistance. But I am digressing, for it is actually the experience of inventing that I am presently trying to weave into our discussion of how the brain works.

Creativity

The mental images of Gentleman Jim's potential mechanism, conjured while stroking a cat, had appeared in my mind as though the components were real objects. Yet they were not. I was manipulating electrical patterns in my brain. It is tempting to call them abstract patterns, but they were not merely abstractions. The various noun-concepts (e.g. brass damper, Plexiglas barrel, air chambers) had resonated together in my head with so much informational depth that they had allowed me to conjure new relationships that might function in the real world—and they had. (I later learned that there are off-the-shelf devices for limiting bounce; it seemed that I might have reinvented the wheel. It turned out that commercial devices generally use the first recoil as the signal to suppress the successors, so they don't generally satisfy the stricter requirement of quick freezing.)

The 'I' Concept Revisited

Shifting attention from engineering to subjective sensation, consider your interpretation of the word, "I."

When you say it, or think it, how much unconscious meaning mushrooms up? I suspect that your entire personal narrative, and perhaps reprocessed versions off much that you have forgotten, constitute the database that each brain accesses under that cue.

A simple form of self-definition comes from this: *I am the process that uses the assembly of synapses that are consciously and subconsciously activated when I think, 'I.' These are the memory record of self, the accumulated past.* So ends a chain of thought begun for me by Karl Pribram, decades ago. Now we can add an important complementary insight. When the thought, *"I can do that,"* arises, the electrical patterns associated with the sense of self have interacted with another set of electrical patterns, the ones that represent the *context* within which the 'I' must behave: the real world. As I've been emphasizing, comprehension of the latter seems to be the province of the silent hemisphere.

The mental experiences that I have described in this chapter are generally flattering to the human psyche and its status sonar. But they don't contradict an earlier fact to which I was trying to draw greater attention: trial and error is full of error, and this is bound to have consequences.

Epilogue

Here is a quick summary of how Plan B came to pass. Because freezing machines had a market limited to the research community, I foresaw that the university patent committee would be unable to decide whether to invest in a patent application. So I published the design and waited for the 12 month time that is allowed for those entitled to patent rights to lodge a claim. As expected, the committee had taken no action, and so the invention became Public Domain.

The significance of this tactic, which was contrary to so much patent folklore, was that the prior rights and normal indecisiveness of committees could no longer restrain me from doing the manufacturing myself—and that is where most of the profit lies. At this point, I was serendipitously offered a tenured Associate Professorship in at the University of Maryland Medical School in Baltimore. I accepted the position so that I could buy a row house with a basement in which I could assemble the instrument on weekends.

Several years later, with about 100 machines sold, the headwaters of the river of bullshit came into view.

I hope you remember about the river? We have arrived in the shallows; there are bulrushes all around. It only remains to recognize the place.

Chapter 19
The Verbal Interpreter

"The shovel is...
for shovelling out the hen house"

Knowledge of the pervasiveness of the trial-and-error principle, as I learned it from the ghost of Herbert Spencer Jennings, was like owning a single chopstick; it was intriguing but somewhat useless on its own. But one day in 1985 I read Michael Gazzaniga's *The Social Brain*. Gazzaniga described an experiment done by himself and Joseph LeDoux with split-brain patients. I'll describe it in detail but first I should refresh your memory.

Split-brain surgery, carried out to relieve intractable epilepsy, has confirmed that only one of our cerebral hemispheres develops prose speech and answers questions. We identify our "selves" with that hemisphere. When we silently rehearse what we are going to do, processing our thoughts at forty bits per second, this is where we sustain the inner monologue (in which we answer our own questions, like a ventriloquist and his dummy).

If a surgeon removes the prose-silent right hemisphere while we are conscious, the knife wielder will generally ask how we feel immediately after the last of the disconnected tissue is dropped into a stainless steel bucket. You'll need a moment to feel aghast.

So far, everyone has replied, "I don't feel any different."

The self simply does not perceive the absence of a prose-silent hemisphere and was probably unaware of its prior presence, (Austin et al., 1974). Consequently, it is just possible that the reality of cerebral duality, which really ought to strike a thunderclap in your mind, and perhaps shift the course of your life, has not yet done so.

A recent clinical case provides up-to-date relevance. In 2008, Senator Ted Kennedy was diagnosed with an aggressive brain tumor. If it had been in the prose-silent hemisphere, the clinical recommendation would have been to remove it entirely, accepting the difficulties that he would then have had in moving the opposite side of his body.

Unfortunately, the tumor was in the verbal hemisphere. If that had been removed, he would have lost the ability to produce speech, a tremendous loss to a politician. Consequently, Kennedy opted for limited surgery that attempted to remove as much of the cancerous tissue as possible, hoping that chemotherapy would keep its further spread under control, while allowing him to cast votes in the Senate. In other words, he opted to retain his sense of self and of context comprehension until he died, which happened in August of 2009.

Although I have alerted you to the context-appreciating function of the hemisphere, we were slow to recognize its role. As Gazzaniga expressed the problem in *The Social Brain*, p.70 (1985), "The right half-brain has no language capacity and along with that a striking inactivity that borders on behavioral tedium. This is not to say that these right hemispheres do not have specialized systems. They may have, but it is next to impossible to demonstrate their existence in a brain system that is so unable to behave overtly."

This statement was later amended for it actually exaggerates the right hemisphere's lack of language capacity—the tissue understands language quite well, but will not speak prose.

The Experiment

Michael Gazzaniga studied at Caltech under Roger Sperry, and participated in the early psychological analyses of the first group of split-brain patients. He then went on to become a neurologist. Though the numbers of such patients remains small, he has continued to investigate another series produced on the East Coast, near Dartmouth. The patients are typically grateful to have been relieved of epilepsy, and normally cooperate willingly in subsequent studies, such as the experiment that delivered the second chopstick.

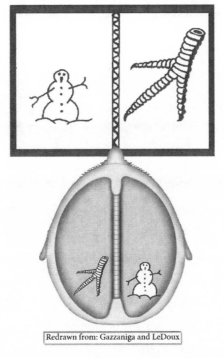

Redrawn from: Gazzaniga and LeDoux

This diagram, (simplified and redrawn from Gazzaniga, 1985), shows the image of a bird's foot that Gazzaniga and LeDoux delivered through one eye to the talking hemisphere of a split-brain patient, while a snow scene was going through the other eye to the silent hemisphere. A divider (the wavy line) keeps the two fields of view separate, and the images end up in the opposite side of the brain.

The most important point is simply that each hemisphere remained ignorant of what the other hemisphere had seen.

The visual restriction was then removed, and the patient 'as a whole' was invited to scan a multiple-choice array of potential correlates. He was asked to point to an image that would be associated with the previous scene. Two responses were expected, for there had been two puzzles.

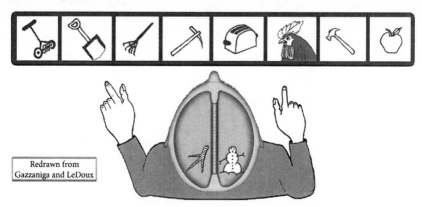

Redrawn from
Gazzaniga and LeDoux

The talking hemisphere pointed to a hen with the hand that it controlled. The silent hemisphere pointed to a shovel with the other hand. If you try the task for yourself, you may notice that snow resonates most readily with the shovel; a bird's foot won't work very well with anything but the hen.

Roger Sperry had already shown that the two hemispheres were independently intelligent, and the above task was very simple, so it was not surprising that each hemisphere pointed correctly. What then was the experimental purpose?

Gazzaniga and LeDoux were interested in how the subjects would *explain* their solutions.

Now is the time to remember that only one hemisphere can talk, and the talking hemisphere saw a bird's clawed foot. We might expect that it would use speech to explain its own logic and it did: it said, *The bird's claw goes with the hen.*

240

But what will the verbal hemisphere say about the correct solution to the other puzzle, that it didn't see? The shovel was pointed to because of its associations with snow—but the talking hemisphere never saw the snow scene.

It will help the reader to have the truth clearly in mind, a truth that the verbal hemisphere might have spoken: *I didn't see the other puzzle, so I'm not sure why my left hand pointed to a shovel.*

The first element of uncomfortable result is this: truth is not the default behavior of the verbal hemisphere. In this instance the patient said, *and the shovel is for shoveling out the henhouse.*

Such false assertions are reproducible: different puzzles result in two accurate selections, but one of the verbal explanations is always false. Different patients make up different, vaguely plausible stories, but they never admit to the fact that they don't have the data from which to draw a conclusion. I am describing the result using language that has pejorative associations, and these may violate the image of the human being that our tiresome status sonar may prefer us to cling to.

Despite that, a fuller analysis of the experiment will show that the language is precisely accurate: it is our instinctively flattering interpretation of ourselves that needs to be reconsidered.

Review

I had become a neuroscientist in hope of comprehending the pervasiveness of bullshit/rationalization. Recognizing the problem early, and amassing a hooraw's nest of diverse examples, I suspected that something about the cerebral dichotomy was important to the eventual answer.

I had also learned that trial and error behavior is perfectly practical, although most people seem to feel demeaned by the idea, and don't want to talk about it.

These elements were provocative but they had not gelled into a complete hypothesis until Gazzaniga and Le Doux's experiment revealed that thinking itself remained rooted in trial and error. It was now rational to suggest that, as far as we presently know, we are inheritors of the most advanced rationality engine in the cosmos. But it begins with trial and error, and we are misusing it.

A Personal Reaction

I read the patient's assertion that the shovel was for shoveling out the hen house while I was sitting in Louie's Cafe Bookstore in downtown Baltimore one Sunday morning. As I started to laugh the laugh of surprise, a wave of nausea hit me. I realized that it was time to put up or shut up: *If isolated left hemispheres spout nonsense as though it were truth, the source of bullshit is plain. It only remains to put the issue in perspective; to explain how we ever get anything right; a task to which I might be suited.* This book and its follow-up are the long-delayed attempt to fulfill those ambitions; they attempt to both name the beast—and wring its neck.

I am not alone in ascribing deep significance to the result. In *The Feeling of What Happens,* (1998), Antonio Damasio writes: "Perhaps the most important revelation in split-brain research is precisely this: that the left cerebral hemisphere of humans is prone to verbal narratives that do not necessarily accord with the truth." Damasio's cryptic acknowledgment appeared thirteen years after Gazzaniga's 1985 review in *The Social Brain.* If the only half of the brain that talks is prone to tell narratives "that do not necessarily accord with the truth," we surely have some serious implications to consider.

In *The Accidental Brain,* (2007), an excellent primer on the nervous system, David Linden supposes that the left hemisphere's potential to come up with untruthful narratives might explain conflicting religious beliefs. I agree. However, 2007 was twenty-two years after the experiment had become widely known, so Damasio and Linden's remarks suggest that many in the neuroscience community have been feeling the weight of this unflattering and portentous result, on their relatively silent hands. Perhaps we have all found it difficult to state the blunt simplicity: humanity is as inescapably reliant on trial-and-error thinking as the microscopic infusorians are still dependent on trial-and-error behavior.

Some of this reluctance to be human critical inevitably arises from our status sensitivity. Another part may reflect the focus of medical school neuroscience on overt pathology. As much Buddhist literature implies, however, 'normal obsessions' threaten our species, and an explanation of the underlying neuroscience could buttress species survival. Interestingly, the Buddhist approach encourages practices that dampen left hemisphericity and augment the activity of the right hemisphere. These are obviously apt if they work, and they do seem to work.

A Choice of Interpretations

While reading Michael Gazzaniga's review of psychological research in *Human: The science behind what makes us unique* (2008), I found this assertion: "Though it can make mistakes, the Interpreter is usually right." The statement is made with little qualification and without reference to any evidence, which is curious because the bibliography is quite large. I must disagree. I believe that the Verbal Interpreter is an expression of the ancient tactic of trial and error, and that it is usually wrong, as all such systems are likely to be. There is primary evidence for this point in the subjective experience of one's inner dialogue. It will be crucial to win the reader's understanding here and so my inner dialogue is laid bare in the next chapter.

Chapter 20
Testimony of Introspection

The Internal Dialogue

James Joyce introduced a literary style called 'stream of consciousness.' When critics claimed that it was artificial, Joyce counterclaimed that it better represented what happens in the mind, that all brains experience a jumble of loose associations that constitute a stream-of-consciousness-dialogue. He asserted that standard literary exposition is an artifice reworked for clarity. He was correct; his critics were full of you know what.

Have you ever silently rehearsed what you should have said in a given situation? If so, we can study the inner dialogue together. I'll testify to specifics, attempting to describe an aspect that the reader's emotions may initially seek to deny. When my cat doesn't come home, I start to experience possible verbal explanations in my head. When the phone rings, I wonder who it might be, and an inner voice often makes suggestions.

Those are fairly low-key examples of mental wonderings. When a person I know seems to look in my direction and frown, however, my verbal hemisphere goes haywire with potential verbal explanations. I'm fairly sure all that is common, and I take it to be the Verbal Interpreter's work in my left hemisphere, providing interpretations, as a good interpreter should. Here is where things might get sticky. I've checked the accuracy of that silent verbal traffic: I have mostly worried about things that didn't happen. The voice on the other end of the phone line is rarely the one I guessed at. Like the bottomless duck pond, most of my ruminations turn out to be incorrect. I assume this is also normal experience, and I hope you agree. I'm afraid, however, the reality is still worse. My first response-thought is often so wrong that I might be considered unemployable if it were part of my formal record. This is the level of honesty that usually puts me out on the end of a yardarm, for the notion of being reflexively wrong most of the time is in conflict with our inherited and pervasive status sonar.

Nevertheless, here's how it goes with me: when mental errors become conscious I feel a sensation of disapproval, an internal wince, to which I often respond by formulating silent rejections: *That's silly!... That's stupid!... That can't be right!* If my random thoughts seem bad enough, they may be accompanied by a clenching of the left side of my jaw, and a piratical closing of the left eye. Is an asymmetric scowl a coincidence, or does it reflect right hemispheric wisdom as it rejects error? (I don't know.) I certainly don't articulate these internally disapproved notions to anyone else. And so the high error rate of the inner voice has no ego consequence. That was how I managed to have a career. Do I ever get anything right? Occasionally, an interpretation is met by a warm glow of approval. This provokes a different response: *I like that!... Of course!... That's nice!.... That must be true!* (The latter is very treacherous.)

Once a thought meets with a positive internal response, I may verbalize it for the benefit of the outside world. When I do so, it binds to my ego. It becomes part of my psychological history; it is incorporated into an estimate of my social status. (Composing this book was strugglesome because of awareness that I would be stuck with whatever I wrote.)

Unfortunately, conflicting information may well show that 'I' am still wrong. Self-correction at that point will involve shame. I would like to assume that these subjective experiences are also normal.

I should note that the presence of internal errors doesn't actually bother me in the long run, for I regard them as a price worth paying for having any original ideas at all. I suspect that freedom from an obsessive concern for 'rightness' is the paradoxical source of an intellectual advantage.

Summary: The error-ridden dialogue that supports one's inner soap opera is the normal correlate of the behavior Gazzaniga and LeDoux's experiment revealed in split-brain patients. Once one accepts that the thoughts reflect standard trial and error, it is not hard to guess at a further strand in the evolutionary design logic of separate cerebral hemispheres.

In one hemisphere, remembered personal experiences provide a reference base that can be reviewed indefinitely in a search for some resonant overlap with the present. As loose associations are discovered, humans turn them into *verbal* snippets of cause-and-effect, plausible logic. But large numbers of possible correlations will turn up, and most of them will be destined for the error category, so the brain as a whole is obliged to review and reject a great deal of what the left hemisphere consciously 'thinks.' There is a corresponding bumper sticker that I can endorse:

Don't believe everything you think!

My inner mental life includes an impression that spontaneous verbal thoughts are immediately sent 'somewhere' for broader evaluation. The Cinderella hemisphere on the right, silently hanging around and much shunned—but comprehending the context—is the obvious candidate place.

Her usual reply is an emotional thumb down. In either case, however, my verbal self decorates the response with more words, typically the phrases I've cited on the previous page. This picture of cerebral function implies a frequent hemispheric alternation of emphasis: a testable tick-tock quality. This seems to be visible in the phenomenon of Binocular Rivalry, which we will come to later.

Bearing Witness

My attempts to convince audiences that inner errors reveal the Verbal Interpreter's presence in their lives have been problematic. As I sketch a mental life full of internally loose associations, some of which would be criminal offenses if acted out, a tension seems to spread through the room. I enjoy the theatrics of doing this, so I probably overdo it. Nevertheless, I am struck by the absence of affirmative nods, the sudden frost in the air, the widened eyes, and the facial stiffening. I seem to have entered territory that is taboo. Although I do worry that I might provoke someone into an epileptic episode, I feel that this taboo is a social Humpty Dumpty that needs to be pushed off the wall, left shattered beyond reconstitution.

On one of these occasions, a Machiavellian strategy came to mind and I caught the wave. Since no hands were being raised, I feigned surprise that every one of my unresponsive audience's silent thoughts must prove to be a correct interpretation, in no need of rejection. I expressed amazement at being in the company of such concentrated genius.

I elaborated further, effusively suggesting that the problem must have been all mine; that I obviously had a uniquely stupid Verbal Interpreter! Feigning shame, I suggested that the wrong person was at the podium. The horrible thought that I might now invite someone to replace me seemed to work: shy smiles became evident, a little hint that an often-wrong internal theorizer is also present in my brothers and sisters.

But the admission seems to be as welcome as a revelation that there is a pedophile in the family. Alice Flaherty, author of *The Midnight Disease*, (2005) quotes a Cajun expression which is relevant here, *If you ain't talkin' to yourself, you talkin' to the wrong people!* I am adding that I know that you know that most of what you say to yourself is highly questionable—it's mostly bullshit!

So now I am challenging the reader, distant in space and time: do you feel any personal relevance in the demonstration of a Verbal Interpreter function? In other words, does the notion of a wellspring of loose associations that are often turned into dubious rationalizations that must be quickly discarded apply even to you?

Acknowledgement may demand a significant alteration in self-image, a self-critical change that is difficult; we would rather not. But we humans suffer terribly from the negative consequences of literally self-serving habits of thought. To overcome this ever-present facet of thinking, we have to face it, understand it, and use it more carefully.

Pause to Reload

I may be prejudicing my case by use of the word 'bullshit.' I once suggested that if we are made in the image of God, that the deity is the biggest … well you can fill in the rest. Lately, I've tried to develop a non-pejorative approach.

How's this: The left hemisphere pops verbalized resonances like popcorn. They fire unpredictably from the popper, driven by loose association, not validity. A shovel for shoveling out a hen house might have been valid. By throwing up enough loose associations one might find a correct one.

Nevertheless, rejection of an enormous number of errors will be required. Since resistance to this admission seems based on misplaced pride, let me offer up a gold standard of thoughtfulness for consideration: How many errors did Einstein reject before the malleability of space and time occurred to him? When we think of Einstein as being amazingly right, we have glossed over the number of times he was willing to think, *Das ist nicht richtig!* and go back to the beginning. Intellectual success is ninety-nine per cent error-rejecting perspiration.

The Crossword Puzzle

The crossword puzzle seems to be instinctively based upon the evolved relationship between our hemispheres. To make this analogy clear, I must first address trivially right answers. If the first clue is: *name of first man,* and there is space available for four letters, most Christians would get the answer immediately—but this illustrates memory retrieval rather than ruminative thinking. Similarly, many crossword puzzle experts will be immediately right about clues and answers that they merely remember from previous crosswords. Crossword puzzles only become analogous to hemispheric management when an ambiguous verbal clue invites the left hemisphere to deduce a word, and the context is a known number of empty letter spaces.

When I recently challenged my thought process to come up with the name for a bird that had four letters, my first inner response was bat. Laugh at my stupidity for a moment. Then more deliberately notice that the loose association phase of thought had delivered two obvious mistakes.

A bat is a flying mammal, and its name has three letters—so it didn't fit any of the clue's criteria. Perhaps the dead bat on my doorstep that morning had left its resonant trail in my brain. Of course I rejected this response swiftly, but in the context of introspection about thinking, it needs to be noticed. It illustrates how the first loose associations of the ruminative process often have to be quickly discarded. After a moment of chagrin, my inner mechanism generated wren, fowl, duck and swan. In between these four, there also seemed to be subliminal errors that I rejected before becoming fully cognizant of them.

Now for a philosopher's point: even if one of the above conscious four was eventually found to be correct, the net success ratio would have been one success for five trials (counting 'bat'). Once again, that is what I mean by saying that thinking involves more errors than successful trials.

Once we find a clue-related word with the right number of letters, we then bring in still larger context consideration, for the answer has an obligation to fit the context of the letters of any vertical answers that are already in place. More rejections, more errors, are usually required. We might find that two o's seem needed in the above bird name. Coot or dodo might erupt from the mental machinery. Once again, a context check will tell us exactly where the o's need to be. Eventually we get it right, and the final appropriateness of the fit is often a source of pleasure and pride.

In the crossword analogy, it is easy to see that becoming right involves many instances of correcting errors—by multiple oscillations between detail and context considerations. The specialized skills of bilateral hemispheres seem to reflect evolution's discovery that, in the absence of occult sources of higher knowledge (that might feed us instant 'right' answers) trial and error is a fundamentally practical strategy.

250

The Wages of Honesty

Gazzaniga's laboratory demonstration of the Interpreter's mistakes resonated with my mental impression of many personal errors, and I hope it does with yours. When I finally connected this with my memory of Jennings' observation of trial and error in single-celled organisms, I heard the 'kaching' of verbal epiphany: the bullshit that emanates from the isolated Verbal Interpreter is clear evidence that we are still trial and error machines!

Gazzaniga Reconsidered

Now let's go back over Gazzaniga' and LeDoux's experiment with resonance in mind. The bird claw image that first went to the verbal hemisphere would have inevitably resonated with whole bird memories. The hen seen in the multiple-choice array would have resonated with and amplified those bird memories. The subject equated a head full of resonance with "finding" his answer. No problem here.

But remember the mistaken assertion, *The shovel is for shoveling out the hen house.* How could resonance explain that? Many people chose this point to take one last shot at denial, suggesting that the shovel might have been useful for shoveling out a henhouse. Why call it a mistake? The answer is rather compelling.

Right after split-brain subjects heard themselves utter the wrong explanations, an evident wave of embarrassment led them to make up excuses for not continuing. For one young man, Gazzaniga stopped the work and discussed how they were both aware that the patient's brain was split, so that some tasks might result in hand movements that he couldn't consciously explain. The patient replied that he knew that—he had said so to his girlfriend on their last date! Cajoled back into the experimental setup, however, an uncontrollable reflex seemed to compel a premature assertion.

Once he heard himself advancing another falsehood, the subject again wanted to stop the experiment. It seems that the silent hemisphere is not satisfied if an explanation <u>might</u> have been valid; if it knows the truth, it wants to hear truth explicitly expressed. When it hears instead what I am calling an error, it generates a wave of shame.

In the shovel case, the subject's verbal brain probably attempted to overlap the two items it knew about: shovels and hens. It is hardly surprising that no resonance was found; I don't think one can shovel live hens; the idea will not resonate with reality.

The verbal hemisphere then seems to have gone looking for older memories, and the mess on hen house floors turned up. That mess *would* resonate with a shovel. Nevertheless, merely seeking and finding resonance doesn't guarantee accuracy. The subject had produced, instead, a rationalization.

Summary Statement: The verbal hemisphere can search the narrative of self, one's life experiences, using any given category heading, until it generates positive feedback between loose associations. It has the further skill of being able to verbalize these correlations as superficially logical sentences. In the unique circumstance where the brain has been split, the Verbal Interpreter immediately articulates the first of these—whether accurate or not—with no restraint. Once the silent side of the brain hears the Interpreter's error, however, it mounts a belated restraint that registers as embarrassment.

Howlers from School Children

It is not necessary to split the corpus callosum to collect hard evidence that loose associations are generated in the human brain and converted into verbal language. Before they become self-conscious about being wrong, younger schoolchildren are a delightfully innocent source. The following examples show the reflex at work.

Who was Joan of Arc? *Joan of Arc was a good woman. She was Noah's wife.*

If no prior associations turn up, an attempt to use logic seems to create them:

Who went into the Lion's Den and came out alive? *A lion.*

The Mental Pillbox

Previous chapters have betrayed a characteristic pattern of verbal aggression and defense; here is a short list of reminders. The street cry: *What are you going to do about?* The competitive academic's tearoom sneer: *Are you serious?* The established professional's plea: *Don't rock the boat when you don't have all the facts!*

These may illustrate the left hemisphere's obsession with short-term logistics of survival and status protection, and this compulsion may explain why so many of us are reluctant to admit to inner trial and error. We prefer to leave a vague impression that something rather more admirable happens inside our minds.

Some of us may also have been so traumatized by childhood chastisement for being wrong that we now conceal from ourselves how many mental errors are necessary to get anything right. Perhaps we have responded by making few attempts at innovative thinking. For such individuals it may be liberating to learn that valid insight is only possible after stomping through a mire of dead ends.

Exceptions sometimes illuminate the rule. There is an implicit admission of our propensity for error when we are trying to be creative in a group setting and admonish each other to refrain from criticism so that ego-sensitivity will not exert prior restraint. But the default expectation is one of a need to defend status. Moreover, when a creative job is done, when it is time to write up the story for publication, it becomes recast as a noble quest from one milepost of logical deduction directly to the next. It is far easier to understand such retroactively logical narratives; they are more efficient communicating and teaching tools. Constant reliance upon seeming to be logical, however, and the social status that comes with that impression, obscures the work of trial and error. Refusal to give due weight to this possibility may block further insight into the hemispheric relationship.

Indirect Support

In *Phantoms in the Brain*, (1998) V. S. Ramachandran writes that neuroscience is still in the 'Faraday stage' when tinkering—merely exploring the range of phenomena that are associated with a certain issue—is appropriate. I agree. If the brain's trial and error philosophy is as basic as I suspect it to be, however, then tinkering never stops; subsequent theory-driven investigation is merely a change of scale that has been decorated with adjectives that are more pleasing to status sensitivities.

Epilogue

During the Reagan years, funding for medical research was deemphasized in favor of the military industrial complex. I lost my NIH grant and carried my laboratory forward by studying the lingering toxic effects of nerve gases, an effort for which the whole department was then being funded.

The army's barely concealed actual interest was the hope of developing a battlefield antidote to nerve gas exposure. Although our research raised doubts about the practicality of this idea, our collective evidence was ignored and pyridostigmine was administered anyway to U.S. Troops in the first Gulf War.

Its daily use on a whether-you-need-it-or-not basis has now been acknowledged as a likely contributor to the 'Gulf War Syndrome.' Even though we had been against it, our efforts had been folded into the great military-industrial-academic bullshit machine.

Although academia had also delivered the necessary insights into my search for the origin of bullshit, my sense of career satisfaction was now in negative territory. Nor could I immediately see how to produce a coherent written account of where we were going wrong. I needed something like a Bodhi tree to sit beneath while I wove many strands of thought together. If I succeeded, I would then need what Archimedes alluded to when he discussed the principle of levers, "If I had a place to stand and a long enough lever, I could move the world." It was time, once more, to review my assets.

My academic salary had generated a pension fund that could be turned into an annuity whenever I chose. Moreover, *Gentleman Jim* had left me with a hundred and twenty-five thousand dollars in the bank. I decided to resign and establish a place of personal sovereignty, someplace pleasant where I could at least sit and think.

Douglas Fir became my tree, from the wood of which I built a home on an island in the Pacific Northwest. My lever seems to be the emergent technologies of the Internet and self-publishing.

Before I go on to describe that phase of my life, and the recent research that has solidified my thinking, I'll conclude Part One with another controversial lecture.

Chapter 21
Dysfunctional Behavior

Learning to lecture

The primary teaching obligation of our small pharmacology department at Northwestern was to medical students. In the context of central nervous system pharmacology, they needed to know whose behavior was strange enough to warrant drug therapy, and what the options were. Chairman Webster gave me a sort of free pass for the first year, expecting me to discover and recover from rough spots, but he made it clear that student comments would be part of the criteria determining whether my three-year contract would be renewed. Since I already suspected that civilization itself was dysfunctional, my task seemed rather delicate, and I plotted a safe set of first-time lectures.

Les had a no-nonsense approach, in which we all got to see the criticisms that students had provided about everybody else, including Les himself. The result was a tall stack, in which I received a discouraging net evaluation of "About Average." I found that the specific critiques, however, were apt and incisive. Moreover, there were consistencies in the criticisms of other lecturers that broadened my sensitivity to what wasn't helpful to our tuition-paying customers. As a result, I decided that the standard procedure of being safe but boring would have to give way to a more personal style. I relegated the dry lists of symptoms, treatments and side effects to lecture notes to be memorized for the exams. That gave me the freedom to develop my own view of brain and behavior during lecture time.

I had noticed something else that became important. When clinicians were able to bring live patients into class, there was a decisive mood shift. I now see this as encouragement of empathetic, right hemispheric participation in the classroom experience. Without any such understanding then, I wondered if I could use my repertoire of personal stories to make a clinical case out of myself. If so, it would be very convenient, for I turned up wherever I was. I resolved to try, and so began my second lecture series with pharmacological blasphemy: an assertion that there are dysfunctional behaviors for which we have no drug treatment. I chose racist thinking as my example.

Novelty Drives the Verbal Interpreter

I was three or four. Mother and I had ridden a double-decker bus past brick-walled warehouses pockmarked with shrapnel damage from the recent war. We were going into Liverpool, to the fish markets. Unknown to me, there were cellars beneath our feet, still fitted with manacles for holding African men and women awaiting horrors of transport only a century or so earlier.

He was as black as coal. I must have stared, before finally tugging on mother's hand. As she lowered her head, I presented my instinctive rationalization: *"Does that man never wash himself?"* If racism is derogatory rationalization, this was it. (I am indebted to a reviewer who pointed out that my family was inevitably taking coal deliveries from coal men in those days. They were often covered in black coal dust—my probable source of meaning.)

As I told this story in the lecture hall, with black students present, it caught everybody's attention; this was clearly not pharmacology as usual. Before I went on to make the point that my question had been an innocent loose association, i.e., a fundamental behavior of the brain and generated without ulterior motive, a young woman in the middle of the class put up her hand. Fully aware that she might be feeling a compulsion to 'put me down' and believing that she had every right to try, I invited her input. We learned that, as a toddler, opening the front door at the ring of the bell, her first racial encounter had been with an African American postman. After staring in amazement, she said, *"Ooh... You look like chocolate ice cream!"*

My spontaneous collaborator had strengthened the intended point: we now had contrasting loose associations, one spun into a negative and one into a positive rationalization, but both coming from innocent children who had each done nothing more evil than express the inner dialogue. As I summarized the situation in those terms, tension resolved and was replaced by something like shared elation. The class seemed to recognize that we weren't trying to learn from a book, nor were we practicing racism; we were genuinely trying to think and learn together.

In each subsequent lecture, I reached for personal relevance; told the tale; and the class responded with their wealth of human experiences and independent intelligence.

Regretfully, I didn't keep track of names, but those occasions triggered many of the thoughts that have ended up in this book.

Reactions to an Idiosyncratic Teacher

After my fourth year of learning from the students at Northwestern's private university, I was nominated for the Teacher of the Year Award, but by then I was on my way to the University of Maryland State Medical School in Baltimore. In this more conservative environment, my frank approach rang alarm bells. My new dean was a tall, old school physician, evidently proud of his dignified bearing. He liked to point out that tenured basic science professors had received a guarantee of a lifetime's employment; he seemed to believe that this incurred some sort of reciprocal gratitude, which meant to me that he had missed the point. I was confident that I had earned my position, and was an asset to the system. I wasn't so sure that he was.

It must have been hard to reconcile a summary of what I said in class with what the dean understood of what I was supposed to be teaching. On the other hand, the lectures were taped. He could have listened and educated himself. Instead, a memo was circulated in which the faculty was admonished to refrain from making racist or sexist allusions in lectures. Comfortable with the substance of what I had said, I didn't realize that he meant me. However, the same memo arrived at the same time (right after my lectures) in my second year at Maryland.

As I saw the pattern, the angry iconoclast welled up within, and resolved to again take the bull by the horns. At first, I wondered who had complained. Over the years, black students in my classes had been uniformly gratified that I had treated racism bluntly. Nevertheless, large classes include a wide spectrum of personalities, and it is certainly reasonable to see my persona as an implicit challenge to assumptions of unwarranted status; I intend it to be. I concluded that I had succeeded, and that one of those students—perhaps the one who listened to those frank exchanges with mouth-breathing disbelief—had run to his "proper authority figure" to complain about my presentation.

The memos had also mentioned sexual issues. It was not hard to see how something similar could have happened there. In one of the elective classes that I held, discussions of sexual fantasies had arisen in the context of hemispheric laterality. (One side being willing to indulge in sexual fantasies, while the other has trouble even talking about it.) Feminism was in vogue, and some of my many female students introduced politically correct criticisms of *Playboy* magazine, alleging that the women therein were being exploited. I had commented that the smiles on their faces usually seemed quite genuine. Going further, I noted that achieving orgasm, at least for me, is often associated with visualizing such beauty in the mind's eye. I challenged the conviction that those images are intrinsically inappropriate. Pongo must have been spinning in his grave.

As is my wont, I had even pressed the point by asking the obvious next question: I wondered whether women have visual fantasies of naked men, as they seek to reach orgasm? One young lady offered the reply that no, women didn't fantasize at that moment. She declared that they just experience a "feeling" that led to orgasm. I thought this was eminently doubtful but let the class decide whether to end the discussion there or to pursue it. They let it drop; we moved on.

A drop-dead-gorgeous young woman came up to me afterward, to tell me that her boyfriend didn't "like" *Playboy* images. Even my level of devotion to frankness has its limits, and I just accepted her statement without pointing out the irony. (Privately I wanted to excommunicate her boyfriend from the men's club.)

Once again, the problem in situations like my conflict with the dean appears to arise from different context assumptions, and I think the following sequence creates the relevant metaphor.

Our evolutionary history is one of being dominated by a chain of silverbacks. Every last one of us began our personal life as a powerless individual surrounded by dominance. As a result of that multilevel heritage of someone else being in charge, I suspect that many of us can feel sympathy for the dean in the above situation. Getting right down to it, even I can feel an instinctive inner murmur: *Somebody needs to be in charge.*

The difficult qualifying reality is that all the training in thinking that I had accumulated constituted the process of making it appropriate for me to be in charge of my lecture content, regardless of the discomfort my style caused to a few inflexible students. I even had tenure with which to stiffen my ability to practice that academic freedom.

Now let me use some related specifics that may clinch my argument for any reader that remains on the fence. While I was inviting the above discussions, the epidemiology and pathology departments were attempting to educate our prospective doctors about sexually transmitted diseases, for the HIV epidemic was just beginning. The necessary level of enlightenment can't be achieved without breaking down barriers of linguistic coyness about sexual matters. After many puzzled years, they had decided to start by showing pornographic movies to the whole class.

It was always a well-attended lecture. After seeing couples of many races in various sex combinations, exploiting various orifices, the typical naiveties of many young men and women—all the deficiencies that parents had left in their sexual knowledge—were erased in an hour. Once the students had a thorough understanding of the varieties of ways in which the sensitivities of our genitalia are enjoyed, they were further taught to recognize pathological abnormalities. My claim, then, is that discussions in my classes were well within the spectrum of what was already appropriately happening in this medical school.

I found one of my student's term paper particularly memorable in this regard. He was one of those occasional individuals who are unaware of ever having a dream, and our class discussion had left him feeling deprived. For the paper, I had asked for some evidence of using both hemispheres. He decided to try to force himself to have and to remember a dream. His technique was to drink a quart of soft-drink just before he went to bed, and it worked: he woke up after about ninety minutes, feeling bladder pressure, and in the midst of the following drama.

He had thrown a clandestine party, while his parents were away. Unfortunately, theirs was a richly furnished home, and one of the female guests had begun dancing on a glass coffee table—that had broken underneath her. Our student hero awoke with a sense of mortification and of not knowing what he would tell his parents when they returned.

As he wrote all the dream details down on his term-paper-to-be, he had the feeling that he knew the coffee-table dancer. By the time he finished writing, he had recognized with whom she had resonated. A few days earlier, in a small group session, he had been learning about breast exams. The female instructor, herself a doctor, had invited him to palpate her breast, with several other students looking on. He knew that he had been embarrassed in real life, and an alternate scenario had been loosely assembled in his dreaming brain. Freud would have loved it, and so did I.

Again, my teaching was well within the framework of a medical school education. There was a problem. It was not I. It was the status assumptions of the dean.

He was using a frame of reference in which his subjective judgments should automatically be dominant over faculty members that, in the privacy of his mind, he supposed to be subordinate entities. But society had long ago attempted to alter that default attitude, and I understood what society had provided for; it was only necessary to take a deep breath and deploy its provisions.

I chose the timing carefully, and waited for a full year. After completing my third presentation of racism-as-dysfunctional-behavior, I described the memos from the dean into the microphone, for the ubiquitous tape recorders. Speaking as evenly as I could, I stated that if those admonitions were intended for me, it needed to be known that my academic freedom to choose relevant illustrations would not be subjected to prior restraint. I added that I was more than willing to debate this issue in public, preferably before the faculty senate. (National news outlets would have been fine, too.)

This is what is supposed to happen, particularly in universities as opposed to trade schools. Conservative fear is not supposed to impose pseudo-tranquility anymore than loyalty to royalty should have kept the American patriots from declaring independence. The boat is supposed to rock. To see that, one needs an epiphany. To have an epiphany one needs to stop reciting safe verbal mantras. One must cross the axonal bridge to the context-wise side of the brain, where the view is more panoramic. What happened in this case was what I had expected: nothing. Furthermore, the memos stopped. There had been no intellectual force behind the dean's expectation of easy dominance.

Any chance of having a quiet chat with him, of course, was forsaken; his blood pressure would palpably rise if we happened to share the same elevator. Some might say that I "should" have first tried for a quiet discussion. In my view of history, however, the track record of quiet discussions in the symbol-strewn offices of the artificially dominant offers no encouragement. Since ancient times, enlightened societies have wisely arranged for verbal dominance contests in fully public daylight. I am, however, referring to real debate, not the silly one-liners into which so much of our present media discussions have degenerated.

Epilogue

For about forty years, the CBS Program 60 Minutes has been exposing criminal and dysfunctional rationalization on a one-at-a-time, retail basis.

Like King Canute, however, the team has found that it cannot turn back the tide and our situation gets progressively worse. We need a wholesale solution.

In advocating for the Golden Rule, and love of thy neighbor, the Prophets seemed to have that intent. Yet the Golden Rule has now been the standard recommendation for even longer than 60 Minutes; it hasn't worked either.

In the hope of understanding our situation better and deducing an effective corrective, we have completed Part One, a tour of the view from my left hemisphere. Early feedback reports that this is (almost) as broadly accessible as I had hoped, perhaps because each of us is looking out at the world through a similar slit in a similar mental pillbox. Nevertheless, it seems that the human-critical quality of my message does not go down easily and often requires that the book be put aside while its unwelcome implications are considered.

Part Two addresses the right hemisphere's broader perceptions, an evolutionary provision that has the potential to finally raise truth and accuracy above selfish concerns and so permit genuine rationality.

Since we don't directly experience the cognitive efforts of the right side of the brain, however, I cannot generally be as personal in my descriptions and must concentrate on the interpretation of experimental evidence. This is a definite skill and for scientists it is normally acquired in graduate school. Consequently, I shouldn't expect as broad a readership for Part Two; perhaps it should even be presented in a separate book. I've chosen, instead, to keep Part Two attached to Part One, if only because this reflects the actual arrangement of the hemispheres in the brain.

For the generalist reader, I believe that flashes of recognition of personal experience will continue to be informative and worthwhile. I hate to suggest that anyone read my prose repeatedly for that is laborious—but it does help and is, in fact, the technique that I have to use when attempting to digest original research literature.

Part Two

The Silent Witness

Figure: Sunspot by S.P. Langley (1891)
(Doubled and bookmatched)

Chapter 22
The Alanmo

A Log Home?

I sold my row house home in Baltimore, converted my pension fund into an annuity, and headed out to the San Juan Islands, arriving during a stock market crash, in October of 1987. Fortunately, my capital was in cash.

My building site had a southern view across San Juan Valley to the snow capped Olympic Mountains on the far side of Haro Straits. Along those waters, tugs pulled floating log booms. A boom is built by daisy-chaining outermost logs together, forming a rough circle that can be tugged to a lumber mill. If the sea becomes rough, however, an interior log might find itself riding up and over the outer ring. In this manner, winter storms delivered many escapees to the long waterfront of Haro Straits, known as South Beach.

In addition, whole tree cadavers, complete with their root systems, were sometimes brought in on the tide, presumably eroded from neighboring coastlines. Although the surface of the bark-stripped lumber weathered into a dull gray, the wood below was perfectly sound.

At one time, licensed logboys rounded up branded logs and redelivered them for a fee to the originating lumber companies. When I arrived, permits to do this were still available but nobody had done any active collecting for some years and the accumulating logs had become a fire hazard. Rustlers occasionally removed brands from the logs' ends: a simple cross cut with a chainsaw generated a Frisbee-sized castoff, which left the logs ownerless and usable for local construction. Taking full advantage of the general situation, a family living in Seattle had bought land with a cliff waterfront in the 1960s. They came up on weekends, bringing kayaks with them, and lassoed loose logs on the water, towing them to the foot of their own cliff, where they had positioned a winch. Presto, home-building material was on site. They eventually retired to their beautiful log cabin.

These ideas intrigued me. But my building site, in the island interior, was several hundred feet above sea level, so I would obviously need more mechanical help. I converted my Subaru Brat into a log truck by installing a pair of X braces in the bed. Now I could convey one log at a time to the island interior. So far so good. I noticed, however, that the best logs seemed to nestle several coves over from beach parking lot. Since I didn't have a boat, the only way to get to them was by walking a narrow trail. But how could I bring them back?

When I saw a kid's bike with cow-horn handlebars at the local Funk and Junk, loose associations began their mental journey.

I bought the bike and walked it, with the saddle removed, to one of the coves. Near the middle of the target log, I drilled a one-inch hole using a bit and brace, and then fitted this hole around the spike upon which the bike seat had once been mounted. The forward extension of the log passed between the upswept handlebars, nicely preventing any confusion about which way I wanted it to go. Quite soon I was walking my log bike back to the parking lot, bearing a sixteen-footer with no great strain. My satisfaction began to wilt a little as I noticed the park ranger's car pulled up next to mine. "That's very creative," he said, laconically, "but it's illegal." "Why?" said I, having been told that log removal was regarded as fire prevention. "Because you can't use a bicycle on the beach!" It seemed that tire tracks threatened to accelerate erosion of the beach. My environmental sensitivities needed further honing.

Although I did collect more logs directly from the parking area at South Beach, they grew progressively heavier. Once measuring five foot nine and a half, I seemed to be growing shorter. The notion of getting logs turned into a house began to pall; I was making a mistake.

Stick-Building

The better part of wisdom was to switch to traditional framing with cut lumber from Browne's, the local family owned lumberyard. This is sometimes called 'stick-building' because it is quite practical for one man to heft the pieces. Nevertheless, it is much more efficient if there are at least two people who can then place each end simultaneously in its proper location.

The architecture I eventually chose came from a Dover book of craftsman-style homes, and it had a castle-like aura. When I laboriously translated the floor plans into a three-dimensional computer program, however, I missed the design significance of the impossibly complex, multiple mansard roofline. I substituted a simple roof that I could comprehend. The result was that my final perspective sketch had all the visual affect of an overgrown shoebox. I had made another mistake.

I trudged into the coffee shop next morning, feeling discouraged. My friend, Ed Stiles, a professional woodworker, had built a house after the design of a chicken barn he had known in Mill Valley, California. I know it sounds awful, but it had worked well, becoming a comfortable home. Ed suggested adding another half-story of tower room, set into the southwest corner of the shoebox. When I drew in this asymmetry, it made the necessary difference. I now liked it, and submitted the plans to the building department.

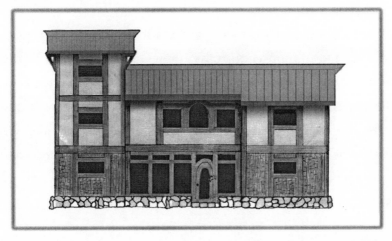

Site preparation was a thousand dollar day. A D-7 Caterpillar tractor scraped the few inches of soil off the house footprint in the morning, and then its bucket was replaced with a massive steel talon that scratched the rock, preparing a trench that would carry water, telephone, and power lines. Next, we tried to form a hole for a 1000-gallon septic tank. (The idea being to sink the septic tank so low that waste is gravity-driven in the intended direction.) The rock in the preferred spot proved harder than the rest, but the tractor did a one armed pushup, with its industrial talon bearing down.

Eventually, a soft spot gave way. All this was accompanied by a roaring that no animal has generated since Tyrannosaurus.

The hole was soon deep enough for a mother D-7 to lay eggs, or for a human to entomb a poop tank. Then I hand dug the septic drain field, for which I was overqualified and too old. I tried to recover while a hired crew built the concrete foundation.

The day after the San Francisco earthquake of November 1989, we were finally ready for the pour. My budget called for me to do all the rest, but it was several years before I really appreciated how long this was going to take.

A Furry Friend

One December, I was having dinner with my neighbors to the North, four folks who had started building their large house after I did, and who had finished it already. A nor'easter was piling snow against the French window where an abandoned grey kitten started crying to be let in. Since I was still sleeping in a cold trailer, I was delighted to have her company. Christened Skypiece, because her fur would have made a nice hat, she quickly figured out that being under the blankets was the best deal for both of us.

I was initially worried that I might crush a small bundle of fur if I were to roll over in the night, but she soon learned to sleep with her paws against my chest. Any unwelcome move on my part deployed a set of little needles just enough to change my unconscious mind; I don't think either of us woke up in the normal course of this reflex, and so we kept each other warm for several winters.

Saturday mornings involved a pilgrimage to garage sales for leftover building materials. One morning I noticed a vinyl record from a band called *Three Dog Night*. I had seen the name before; now I had a resonance to go with it, and belatedly got the point. (As the nights grow colder, one sleeps with extra animals.)

When the 12 x 12 foot tower was framed, one October, God knows which one, I put its tin roof on quite quickly and we were able to move out of the trailer. I hadn't built staircases yet, and so we were reliant upon ladders, one of which had been fabricated entirely from aluminum tubing. Having no safety features, it was very light and readily deployed. Perfect for a circus, or a one-man job site. I placed it against the outside wall of the first floor, so as to access the second.

For several days, the round rungs of this ladder intimidated Skypiece, and I had to carry her up in the evening. One morning I noted a tour de force of cat adaptability: she had figured out how to come down. By reaching both her front paws to a lower rung, she was able to bring down the hind set simultaneously, repeating the double-footed sequence in a slow tick-tock until she was low enough to jump the remaining distance. I don't know how she learned this, but if cats have dreams, and I'm sure they do, it may have come together in that venue. A few days later, while being chased by a feral cat, she added an ability to run up the ladder doing a smooth, single-footed alternation before jumping off near the top. The befuddled pursuer, stuck at ground level, foiled, and presumably gobsmacked in a catty sort of way, could only watch.

As she reached the top, I watched Skypiece turn and look back down. She then slowly walked away from the edge, so that the tip of her lifted tail disappeared gradually from sight. Was her right hemisphere exploiting the context for a moment of haughty disdain? Of course I don't know, but the notion is legitimate; the behavior suggested some such mentation.

Anger Mismanagement

When it was time to sheathe the main roof with metal panels, November storms were threatening. I had nailed skip sheathing to the roof rafters, and had fastened heavy tarpaper onto both south and north faces. Now I just needed to screw down foot-wide panels of galvanized, painted steel, and the building would be weather tight. The panels were all fastened to the south slope just as the first storm hit. But howling winds that night eventually drove horrible flapping noises on the north slope. By morning, tarpaper was all over Dan and Jim's house next door and half my roof was again open to the weather. Repapering it was exasperating, and then I had to lay down those 20-foot long roofing panels that could only be lifted during lulls in the wind. In the middle of this aggravating process, I met a fellow who was having a different difficulty: anger management.

The physical beauty of these islands soothes psychic aches and pains, but the process of packing them with human beings creates tensions. Many retirees have done well in their working lives, have wielded their share of power, and miss it in retirement. Others are working hard to make it possible to stay on the island. Some in each group embrace the authority of neighborhood covenants (rules) with unfortunate passion. To make matter worse, the most neurotic of them volunteer to run the neighborhood associations.

Like the bureaucracies they always dislike, they then fall prey to an instinct to 'do something' about anything. Distressingly often, they misconstrue rules in pursuit of some other agenda. Fairly quickly, all members of the association are being assessed to pay for lawsuits and for insurance to protect board members' personal assets.

The lawyers do well, and then get asked to write the next round of remunerative 'rules.' (Study Question: Does this pattern, like a fractal, continue all the way up to the largest scales of society?)

That's the background, here's the story: Don was living with Elaine, who presided over the neighborhood association in the adjacent neighborhood. They were both walking by and he called up to say that my windblown activities were illegal. When I asked why, he replied, "Your covenants forbid metal siding and the roof is a side of the house!" We exchanged several 'no it's not's' and 'yes it is's' and his constantly ruddy face became still more scarlet.

I had realized that the time it was taking me to complete my project was upsetting some people in the adjacent neighborhood, who would not have allowed that on their lots. When Elaine delivered these complaints, seemingly in hopes that I would respond to them, I practiced what I felt was a Dalai Lama level of calm patience in listening without comment. Eventually she talked her own way through to the point that I wasn't in her neighborhood and so could do as my neighbors allowed. And since most of the adjacent lots were still raw land, I wasn't directly bothering anybody.

Don and Elaine's relationship foundered; he left for parts unknown. Once I had finished the house exterior, Elaine and I related more easily, even dancing together at the local contra dances.

One day, while standing in line at the supermarket, I recognized Don. He looked healthier than I had ever seen him; the flush of high blood pressure apparently under control. We acknowledged each other with a nod, but didn't speak. The next morning, when I drove past on my way to town, I saw police barrier tape all around Elaine's home. Don had shot and killed Elaine and then himself.

The specific rationalizations that had led to this end were never made public.

A Christening

When all the structural work was done, there was still plywood sheathing over the window openings, except for narrow slits that let in daylight. This had given the project a balefully defensive character and I learned it had acquired a nickname: *The Alanmo*. Soon after receiving this more lightly delivered criticism, I decided that the time had come to buy and install windows.

Next I began to notice that the pile of material for the aborted log home, still on the site, was an eyesore. Hearing rumors of a neighborhood petition being mounted to complain, I responded by digging holes beside the driveway, which allowed me to install a sort of wooden honor guard of vertical logs in the ground. The initial visual effect was horrible. Fearing a substitute petition (to re-pile them), I cut the tops down into a swooping curve, from the highest at eight feet to about three feet from the ground. This looked good, and gave another mistaken impression of intelligent foresight. A few days later, I drove home late. It was Skypiece's favorite season of the year, the time of grasshoppers; a time when she fed herself by hunting, and ignored the store-bought chow. At the sound of my car, she jumped onto the three-foot log, and then leapt from one to another until she was looking down on me from high in the air. A full moon completed the scene. I felt the hormonal flush of deep satisfaction and chose the moment to reflect. I had reached my sixth decade and the whiskey hadn't done me in. Moreover, I now had a place to stand.

When the building department signed off on the *Alanmo,* I got sozzled on two pints of Guinness and recalled the time mother had told me to go away and never come back.

Had that experience been a contributing factor? Any link is surely not simple, but any assertion that there had been no connection would be even less accurate, for all our present day behaviors reflect integration of all that went before.

The Mental Swimming Pool

As I turned to the long-postponed book, still just an obsession with Gazzaniga and LeDoux's experimental result, I pondered the expected resistance to my inevitable criticisms of humanity. I wondered how that might be overcome. I decided to trust to the device that I had found so helpful in lecturing: I would be personal to the edge of impropriety.

After several aborted beginnings, I realized that the socially important insights can be conveyed without much jargon. Once I also accepted a need to acknowledge my own mistakes, the text seemed to benefit from synergy between my life's errors (and eventual successes) and the basic thesis.

A further boost to my motivation came from the George Bush and Dick Cheney administration, for they stirred the inner Celt. I began to feel that a combination of personal honesty and blunt prose might serve as a socially acceptable shillelagh.

So now let me share the neuroscience catch-up experience, and then explain our vulnerability to spending a lifetime splashing around in the shallower end of the mental swimming pool.

Epilogue-Review

As a child I learned to my sorrow that so-called human rationality is more typically self-serving rationalization of transitory benefit. As a graduate student, I was astonished to find that self-consciousness is inherently one-sided, and that there was an ongoing mystery of the value and purpose of the silent hemisphere. I suspected that these facts would prove pivotal in any eventual explanation of our tendency to screw up on a global scale.

As I began teaching, Jennings' discovery that trial and error is an ancient and successful foundation for behavior struck me as another fundamental insight. A decade later, I added Gazzaniga and LeDoux's discovery of the left's Verbal Interpreter—and its vulnerability to error. I have here offered a consequent thought that seems to be original but has no right to be: the error-saturated inner dialogue reflects humanity's continued reliance on the trial-and-error principle.

We avoid correspondingly haphazard behavior, however, by applying selection criteria to the thoughts and plans that arise in our inner dialogue. Ideal criteria can be readily imagined: that we would always speak the truth and that we would always compute the wisest action. But, clearly, we don't.

The split-brain experiments established that the left hemisphere readily delivers loose associations that are not accurate. This has led me to finger that tissue as the headwater of the river of my title. Whatever corrective exists must arise elsewhere and must operate by still unknown principles.

But before we turn to the obvious candidate, it is worth noting that bull itself is not criterion-free: it tends to be selected for its self-aggrandizing qualities. There is a ready explanation from evolutionary psychology. Pressure for reproductive opportunity equips social animals with status sonar—and a strong preference for being high in the pecking order. Self-aggrandizing instincts driven from the left brain seem to ensure that this tissue selects those thoughts that may defend or elevate one's status. In contrast to dominance battles, long-term rationality seems to demand a source and mechanism for thought-selection that is more distanced from and more independent of the existential angst of the competitive self

Does the other side of the brain have the wherewithal to fill the need? After the corpus callosum had been split, the silent hemisphere either grimaced or displayed embarrassment when mistaken assertions were uttered by the left. At least to this extent, it seems to have some sensitivity to truth.

In the next chapter, we will further consider the covert character of the truth-sensitive silent hemisphere.

Chapter 23
The Rationality Engine

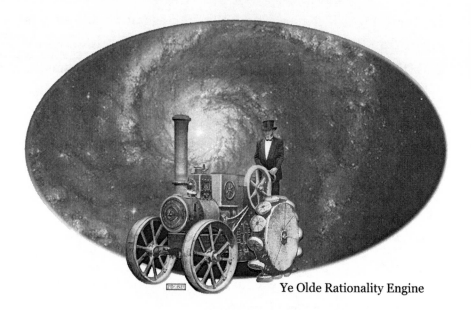

Ye Olde Rationality Engine

The Model of the World in the Lucid Dream

Evidence from EEG monitoring and from blood-flow studies suggest that vivid dreams arise in the right hemisphere and usually go unrecorded by the left's narrative of self. If we awake mid-dream or soon after, however, the self may still register a rapidly dissipating account of the dream.

On the other hand, those dreams that have been termed lucid have such a vivid quality that they even provoke the dampened self of sleep to ask, *"Is this really happening, or am I dreaming?"*

I first read of the phenomenon in a Sunday supplement article about Stephen LaBerge, a sleep researcher, then at Stanford. LaBerge explained that, when the above question arises, it implies a low level of conscious awareness, enough to direct the plot!

LaBerge hoped to teach the techniques to paraplegics so that they could re-experience the existential thrills of skiing, running, even flying, and seduction, (LaBerge, 1990). One night, a week after reading the article, I dreamed that I was skiing along a secluded trail and fell down, landing on my butt. While I was still sitting in the snow, a baby seal slid across the trail and came to rest draped over my thighs. At that point, I wondered if I was dreaming.

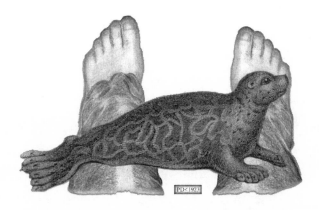

From within the dream, I recollected the news article and accepted the opportunity to fulfill a long-held desire—stroking seal fur while it was still attached to its living owner. I lowered my hands and stroked the seal's back with great pleasure. But the sensory detail was too intense to last long, and I woke up. On other occasions, I have recreated my favorite film star in an evening gown with spaghetti-thin shoulder straps falling from her shoulders. Once again, unfortunately, I become so excited that I awaken before they reach her elbows.

Such effects are reminiscent of mythical tales of Sirens, and of Achilles' dalliance for years in Calypso's cave. These suggest that even in classical antiquity, there was awareness that the world includes shortcuts to such a surfeit of pleasurable sensation that external behavior becomes secondary.

Evolutionary Psychology and Lucid Dreams

What drove the brain to create a mental Garden of Eden, a vulnerability that might bring even the sex drive into a virtual mode? And having found that the capacity to imagine an ideal garden had survival value, how is self-fulfillment within an inner model of reality routinely prevented, so that external behavior is forced to occur in the real world? We have enough information under our belts to offer some coherent responses to those questions.

First, development of two zones of understanding, one for the self and one for the external context, with a controlled potential for interaction, may constitute a net rationality engine of so much survival value that the risks of an internal short circuit were worth taking. The physical separation of two bands of sensory frequencies into paired cerebral swellings may have provided the necessary safety factor. In this view, the inter-cerebral fissure, acting as an anatomical and mental Grand Canyon, may ensure that the self-hemisphere responds preferentially to the Gardens of Reality as they provide sensations to the eyes and ears, rather than indirectly from the Garden of Contextual Enthusiasms. This is reminiscent of the Greek assertion/guess that the left hemisphere senses the world that the right hemisphere better comprehends.

The association of vivid dreams with the right hemisphere implies that it is the right that builds models of reality, within which either the self or an idea might be manipulated and tested. The experience of a lucid dream, however, raises sensations of opportunistic trespass. If the model of reality that they betray is an important component of a mental rationality engine, then the peremptory ejection that plagues mine also suggest that its operation requires that the executive self be generally kept out of the engine room. Relaxed when we sleep, exclusion may be more important when the self is fully conscious.

A clue to an alternate interhemispheric dynamic can be deduced from the testimony of introspection. When a significant new thought arises in my inner dialogue, I usually respond with a question, "Will this work?" or "Is that right?" My internal impression is then of an emotional reply, sometimes experienced as an 'Aha' moment, but one that includes little cognitive detail. This subjectively experienced sequence seems more consistent with the notion that silent questions generated by the conscious self trigger an <u>upload</u> of an idea <u>to</u> the right hemisphere, followed by a pause after which the verdict is delivered back to consciousness. Tentative as such indications are they give us something fruitful to work with, as I'll show in the penultimate chapter. In the meantime, I need to tell you about the genesis of the text vs. context dichotomy.

The Two Sides of Perception

For our fruit-loving ancestors, there was survival value in being able to recognize ripeness from color information, and evolution responded by adding color-sensitive cells to the primate retina. These best detect the frequencies corresponding to red, blue and green light, and a logical recipe in the brain (an algorithm) converts their responses into thousands of hue meanings. With these we still evaluate ripeness as we sort through fruit from grocery-store bins.

In 1998, Richard Ivry and Lyn Robertson's monograph, *The Two Sides of Perception,* introduced a surprising new example of a frequency discrimination. The sensory signals entering the brain are first surveyed and divided so that the high frequency end goes to the talking hemisphere, while the low end goes to the silent one. The result, in the silent hemisphere, is a lower information density, which allows faster processing. This appears to be the physiological basis of an improved description of the cerebral dichotomy: *text* versus *context* discrimination (in left and right hemispheres, respectively).

Consider the folk aphorism: *He couldn't see the forest for the trees.* This refers to a tendency (in some personalities) to concentrate on and overemphasize details—and thereby miss the big picture.

Ivry and Robertson's insight suggests that the idea of a contextual forest emerges quickly from the silent hemisphere's low frequency summary of the general pattern that is created by numerous trees. Yet we remain free to become bogged down in leafy details, to ignore the broad understanding in the right hemispheric unconscious.

What survival value drove the evolution of a double scale of attention? The reality of predatory snakes provides a simple example.

The sinuous S-bend of a snake is an iconic movement pattern that can be instantly discerned. A fast analysis in the right hemisphere might trigger a lifesaving reaction. Some snakes, however, are harmless and would only have survival value as food. A slower but more detailed analysis of the skin patterns, or the head shape, by the hemisphere tuned to higher frequency firings, would enable that discrimination.

A Choice of Dichotomies

Robert Ornstein's widely read book, *The Psychology of Consciousness* (1972) reviewed the 1960s studies on patients with split brains.

I had learned the gist of the message while I was at the Langley Porter Neuropsychiatric Institute, and was well aware that Robert was a careful scientist. In later years, casual conversations with his nonscientist readers, sometimes included claims that "Ornstein believes" something outrageous. I found myself echoing Lloyd Bentsen's political debate line... *I knew John Kennedy personally Senator, and you don't know what you are talking about.* I have also found my own descriptions of the cerebral dichotomy innocently lumped together with tales of psychic parrots. Perhaps this happens because any invitation to open the mind elicits wonder, and so all wondrous possibilities seem relevant.

Although I want my readers to realize that self-understanding has paradigm-shattering implications, I still find parrotic leaps of gullibility hard to deal with.

In a more recent Ornstein book, *The Right Mind* (1997), Robert reviewed twenty-five years of further insights into the hemispheric relationship. He included the evidence that I have delayed relating until now.

A major source of insight into the differing abilities of the two hemispheres arises from neurological observations on stroke victims. We have seen this in the case of Jill Bolte Taylor's left hemispheric stroke, which rendered her reliant on the right hemisphere for all mentation, and which tissue delivered direct context-consciousness on a scale that seemed to embrace the cosmos. What about right hemispheric strokes?

In this case, the still healthy verbal hemisphere continues to perceive details, but is unable to arrange them appropriately. When asked to draw a cube, for example, patients with a damaged right hemisphere may draw four right angles as shown here.

Note that they are nonsensically piled on top of one another. In this case, the functional eye has picked out the horizontal and vertical lines as they meet to define those details that we call 'corners', but the next level of contextual comprehension is absent.

The nearest four corners of a cube belong in a square array, while most views would present three more corners, deeper in the 3-D background—but the patient can no longer download this complex spatial understanding from the damaged right hemisphere.

Expanding Context Comprehension

Conscious awareness, lauded though it certainly is, has a processing speed that can only handle forty bits per second, enough to remember a telephone number as we reach for a phone and dial it, but not much more. This fact may necessitate that context information be held outside of consciousness to be downloaded if and when it becomes relevant. If I imagine the reader sitting at home in a comfortable chair, book in hand, with this sentence being scanned by both eyes, and I bring up your *mailbox*, a sense of the trail between here and there may have just impressed itself on your conscious mind. For a creature that can only do a certain number of conscious actions per second (nothing personal intended), this download-as-needed provision is usually adequate.

Context Expansion During Narrative

As Bartlett showed in analyzing the party game (Chapter 11), when the right hemisphere is available, we instinctively reach for context comprehension first and only then expect details to hang together properly. Once we think we 'see' the broad context, we tend to zoom back down again to the level of current details. The unfolding of a narrative, however, may demand successive increments of broader contextual input.

Further rumination along these lines delivered the epiphany that the experiences that became my collection of stories had driven context revisions as events unfolded. As I finally saw this point, my collection of instances of right hemispheric experiences went from a threadbare troubadour and diver-in-freefall—to being overstuffed with every twist and turn of personal narrative. There remains this difference: during original experience, context expansions do not generally allow sustained 'enjoyment' of the right hemispheric perspective for its message is quickly lodged and one is generally immediately back in left hemispheric executive mode. (In telling the tale later the story can make the most of the context revisions.)

During diving, something I don't do anymore, I usually felt locked into a more prolonged period of right hemispheric spatial analysis (and bodily adjustment) until I hit the water. This was not particularly pleasurable; I seem to have done it for the self-congratulation that came afterwards.

My favorite avenue into the right side of the brain remains immersion in narrative song, for this seems to allow a prolonged experience of empathetic sensitivity, and I think this is what Homer and his audiences were enjoying together thousands of years ago.

Other Animals

There has been a widespread assumption that cerebral lateralization arose in the service of human speech and is relevant to man alone. This instinctive puffery has prevented some scientists from seriously considering that the two hemispheres were distinctly specialized well before speech. A wide variety of evidence now supports the latter view. I particularly like the demonstration that some fish emphasize one eye as they evaluate objects newly placed in their environment. Blind relatives of the same species use the lateral line organ on one side, repeatedly holding that side close to new objects as they swim by. Once the object becomes familiar, however, such asymmetry disappears.

The investigators who made these findings (Sharma et al., 2009) suspect that the mapmaking function (the context analyzer) of these fish brains is lateralized.

An article by MacNeilage et al., (2009) in *Scientific American* reviews the case that cerebral lateralization was in place five hundred million years ago. This bolsters the conclusion that the benefits of big-picture context analysis were recognized early in evolution. Perhaps all creatures with bilateral brains approach reality analysis with these two frames of reference: a small scale within which the self looms large, and a larger scale that perceives the self in context.

This is a good time to reach for diagrammatic equivalents. As shown below, I started by supposing a simple but well-defined context: a rectangular quilt that fills the right side's perception. In the left hemisphere I portrayed some details that might pass through the window slit of conscious perception: the notion that a weaving technology had been at work; that the original thread was white.

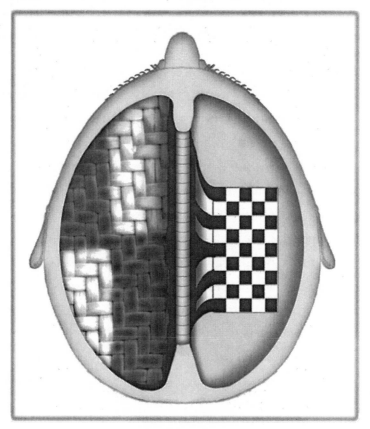

I next considered a cat's brain, below. If it doesn't really look like a cat brain, we'll have to put that down to cut-price foreign labor (me). I assumed that cats would probably lack a word for rectangularity, and supposed that they also lack the concept—so I left the black and white patch amorphous.

Although visual acuity may be high, details of stitching would probably elicit no resonances for a cat, and so would not be attended to. The resulting diagram captures the different scale preferences of the two hemispheres but otherwise depicts simple sensory perceptions and is fairly boring.

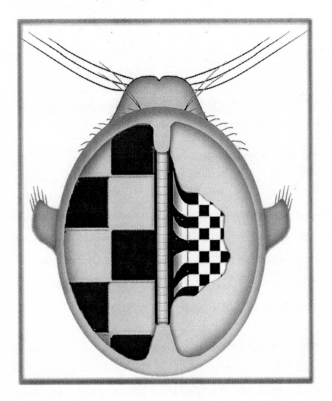

An attempt to empathize with cathood, however, reminded me that cats are predators and so their consciousness might peer into any terrain with a detail concern—that camouflaged prey might be present.

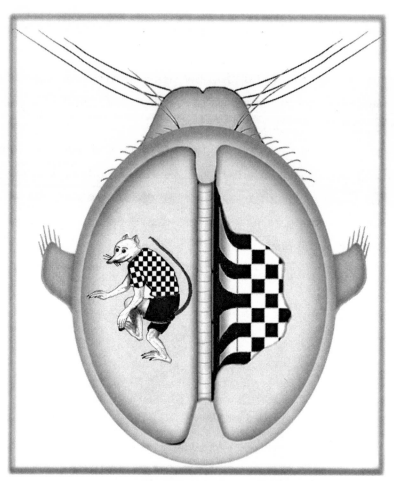

That notion led to the above sister diagram that may capture some of the cognitive differences between the broadly conceiving and the action-oriented hemispheres of another species. If you have the time, this general notion might be an interesting art project.

Contextual Reserves

When I pick up a saltshaker in the local cafe and place it back down, asking what my companions see, most reply, "A saltshaker." Light reflections from the tabletop inevitably impinge on the retina, but the limited information in consciousness seems biased towards details that the self might directly manipulate—the saltshaker rather than the tabletop

In this example, there is also out-of-sight context: the tabletop must have supports that extend to the level of the floor, which is itself supported by hidden joists inserted into the walls of the building, these in their turn resting on foundations set in the earth of the town, county, state, nation, continent, and Planet Earth. Throw in the Milky Way or not. All of this successively enlarging context data is available in your model of reality—but it is held in reserve, outside of immediate consciousness, apparently in the right hemisphere. As each element was named, it became a labeled detail and, as such, briefly passed through consciousness at forty bits per second.

The following metaphor for the situation arose from my time on Lurline. Evolution placed steerage responsibility for a human life in an executive self-hemisphere. The self may fix its gaze on a compass needle, but a silent navigator on the other side of the brain can also hold an encyclopedia of the heavens in reserve.

Multiple Dichotomies

Each of the following word pairs highlight different but non-exclusive facets of the complex hemispheric personality difference.

Word versus picture. This is Shlain's preference in his masterpiece of archaeo-historical analysis, *The Goddess versus the Alphabet* (1998). Imagine a red apple; rotate and even dissect it in your mind. We undoubtedly vary in the degree to which we can do this with any fidelity but that's okay. Now notice that any statement that you make about your apple image will collapse the holistic mental apple into a semantic slice. For example, you might describe the color of the skin, taste of the flesh, location and purpose of the seeds in the core, the curling flakes as you peel it, etc. Each item will be just that: an item. Compared to the flexible set of 'apple' connotations that is, I suspect, imagined visually within the unconscious of the right hemisphere, literal language constructs unwieldy vessels for knowledge.

Linear versus Pattern Analysis. Grammatical language is linear. This may be a profound limit on the left hemisphere's ability to capture multi-dimensional reality. In the silent hemisphere's pattern analysis, every specific seems to remain embedded in a contextual everything else; although how many extra dimensions the right hemisphere can thereby follow is not obvious (at least to me).

Text (or Detail) versus Context Analysis. This arises from Ivry and Roberston's work and has been treated by Ornstein as perhaps the most fruitful dichotomy description in current use.

Literal versus Metaphorical Analysis. This one has merits that are explored in the next chapter.

Binocular Rivalry

If the two sides of the brain have different strengths and weaknesses, partially characterized by each of the above dichotomies, one might expect the left's consciousness to check the survival value of the right's sensations with some regularity. For the simplest consideration, which is direct sensory input, the phenomenon of binocular rivalry seems to show exactly that.

If you can cross your eyes and focus a fused image of the grids in the following diagram, I can demonstrate the effect. The fused image is the one in the middle; it should appear sharp and have pseudostereoscopic depth.

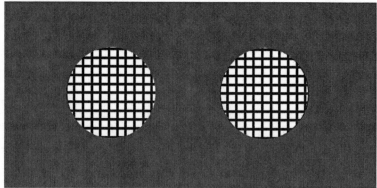

What would consciousness perceive if you could arrange to input something distinctly different through each pupil—let's find out.

With a similar diagram to the above, we can deliberately place vertical lines into the fovea of the right eye by locating them on the left side as shown below. Remember that the right eye is going to be crossed, so it will be pointed leftward. If you are left-handed these details may reverse but the general idea remains valid. As you can also see, the left fovea is going to be offered horizontal lines.

What do you expect will happen? A common general assumption is that the two images will simply fuse in consciousness, so that you will again see orthogonal grids in the middle image. Now is the time to find out. Take your time; independent observation is not as trivial as you might expect.

Many observers report a fleeting impression of an overlapping field of orthogonal lines—but that is not sustained. Instead, the same majority recognizes that their consciousness begins to alternate between an awareness of patches of horizontals, and then verticals, and back again. This will go on indefinitely for the involuntary underlying switch mechanism never gets tired.

This effect seems to arise because of the left's alternation of attention between its own sensory input and that which reaches it indirectly from the right side of the brain.

Can Truth Sensitivity be Downloaded to the Left?

In this chapter, we have noted that both broadened levels of context comprehension as well as indirect sensory information can be smoothly downloaded from the right hemisphere to the left's consciousness. This raises the possibility that the right's greater sensitivity to truth or accuracy might also be downloaded as a bullshit restraint. The well-known 'whisper of conscience' seems to represent exactly this influence. In the working hypothesis that I am trying to build, I hesitate, however, to place much weight on that whisper. For such an important purpose it seems vestigial; perhaps it is a remnant of a previous time. We can at least hope for a more robust alternative.

A second doubt arises as follows. In all probability, your sense of self was in no way diminished when you began to see through the other eye and hemisphere on the previous page. Nor did you experience nirvana along the lines of Jill Bolte Taylor's stroke-induced reliance upon the right brain. In other words, although sensory information traveled through your right hemisphere, its curious personality in no way leaked across into the self-consciousness of the left.

Finally, there is the evidence from lucid dreams that evolution found it necessary to surround the cognitive capabilities of the right hemisphere with a high garden wall. If this is correct, the merely whispering voice of conscience might be considered evidence of chinks in the right's provisions for covert existence. The possibility noted earlier, that ideas are uploaded to the right hemisphere where they are examined in reference to a well-guarded model of reality is for me, the most persuasive.

The Writer and the Editor

Christopher Booker also considered interhemispheric dynamics in his book *The Seven Basic Plots; Why we tell stories* (2003).

Instead of viewing the Muse as the original source of authorial plots, Booker suggests that it merely guides the plot—as an editor. This is closely aligned with the thinking we have explored.

In Booker's version, the left side throws up an initial line of trial-and-error narrative activity. The thought is then passed across the corpus callosum to where the Muse can evaluate it and communicate an editorial judgment—either with emotional enthusiasm, or with repugnance. As every writer knows, many possible thoughts and sentences erupt within, but few get the nod of internal approval.

Booker's review led him to argue that our most durable literature resonates with an iconic pattern: in the beginning of the narrative, a state of happy tranquility is lost under stressful circumstance.

At the core of every plot is the struggle to regain tranquility, usually symbolized by reunion of separated lovers. While that theme is baldly exploited in romance novels, Booker shows how it is also woven into the texture of huge swaths of literature.

What of tragedy? Booker argues that it conforms to the basic plot by painting a grim picture of what happens when successful reunion fails to occur. Booker further suggests that the theme of a struggle to regain tranquility arises because the Greater Self is approving metaphors for an unfortunate parting of hemispheric ways. It wants to get the differently fascinated hemispheric pair rowing again in the same direction; it is striving to harmonize the left's little perspective (on the self) with the right's bigger picture of the self-in-context.

What was the nature of that putative parting? Booker suggests that when our capacity for self-consciousness reached the stage of conceptualizing nakedness, we crossed a Rubicon. Like Julian Jaynes, Booker has sensed something quite literal lurking in the biblical metaphors for gaining self-consciousness. He is implying that language acquisition may have been the straw that broke the back of ancestral innocence.

We met this thought earlier by a different route. Once words like *I, me* and *mine* were introduced, they inevitably amplified any prior nonverbal sense of individuality, and so produced verbal self-consciousness. At that stage language also permitted critical gossip: *Grog's got no clothes on—and a dirty bum, to boot!* As a result, existential sensitivity would have arrived, for the survival instinct would have driven intense discussion concerning death, eventually and reluctantly acknowledging its inevitability.

Booker supposes that the right hemisphere witnessed these seemingly unfortunate aspects of language-focused awareness. In response, it edits our stories, withholding approval until the plot turns on recovery of blissful serenity, or at least conscientious personal hygiene and a happier frame of mind. Booker's analysis of literature thereby comes strikingly close to my conclusions from a life in neuroscience. The biggest difference is this: where Booker sees the hemispheric struggle for mutual harmony as having been instigated by the development of language, I see the struggle as a more ancient attempt to maximize the resonant correspondence between the left's attention to local logic (i.e., cause and effect relationships between categorical details) and the right's broader comprehension of patterns in reality.

Chapter 24
Metaphorical Comprehension

Just as context understanding is vulnerable to right hemispheric damage, so too is metaphorical comprehension; the same patient's speech may become restricted to literal thinking within well-defined noun categories. A complicating factor is that the comfort level with metaphor varies widely, so this particular loss is not always obvious.

Dictionary Definition: A metaphor is a figure of speech in which one object is likened to another by being referred to as if it were the other: *He was a lion in battle.* This statement is literal nonsense: he did not change species' category and become a lion in battle—he was still a man.

What mental process is involved in nevertheless accepting the metaphorical assertion? In the dictionary example, it seems that the man has matched a *pattern* previously associated with a lion savaging its prey. We can generalize from here, noting that metaphors disclose pattern detection operating across different categories of thought.

The picture that then emerges is of the talking hemisphere instinctively partitioning reality by noun categories, about which it can then think using language as an assist. The other side instinctively extracts pattern similarities independent of category, but with which it can also think (perhaps using further pattern analysis). Language assisted thinking has been called discursive. The metaphorical form of thought is called non-discursive.

Another quality betrays metaphor's origins in a part of the brain with a distinct emotional tone: metaphor simultaneously and forcefully fuses thought and empathetic feeling. The result is more sympathetic "meaning" than arises from the mere denotations of literal speech. Perhaps for this reason, exceptional metaphor is the hallmark of great writers and poets—but not of modern politicians.

The Physiology of Metaphor

In *The Political Mind* (2008), cognitive linguist George Lakoff describes metaphor as "a mental process that is independent of language but which can be expressed through language."

He also describes a neural theory of metaphor that emphasizes a modern axiom of neuroscience: *neurons that fire together wire together.* Example: a primary and common sensation, say warmth, may become linked in childhood with specific examples of warmth, such as a mother's affectionate cuddle. That association causes the relevant neurons to fire together, eventually supporting the metaphorical idea of 'emotional warmth.' Note that the child who is held affectionately will already have associated affection and warmth before learning to speak. That is the sense in which metaphor is a mental process based in pattern analysis, and independent of language, but eventually expressible through language.

Metaphorical Reality versus Literal Reality

Jim McCormick, who edited early drafts of this manuscript, spent many years teaching English. He tells me that students vary widely in the ease with which they approach metaphorical understanding. To some, metaphor seems impenetrably foreign, and they resent metaphoric exercises being forced upon them. On the other hand, a writer like Diane Ackerman astonishes me with the rich proliferation of apt metaphoric allusions that she can pack into a paragraph about science. (e.g., *An Alchemy of Mind,* Ackerman, 2004).

298

Rivalry between Literal and Metaphorical

In the last chapter we saw a fluttering alternation of awareness when each retina saw differently oriented lines, direct evidence of alternation in hemispheric attention. Can one set up a situation that might deliver alternation of literal and metaphorical meanings? The following reproduction is of a 150-year-old print from *Les Fleurs Animee* that portrays an herbivore's 'appreciation' of thistles. Thistle spines are literally defensive, but the artist builds a metaphor by portraying the thistle plant as a young woman bent on rejecting an herbivorous suitor.

I cannot immerse myself in a lucid version of metaphor world, but, as I view the above image, my mental seesaw tips in the direction of accepting the metaphorical allusions. In the next moment, however, resistance arises—a feeling that this is stuff-and-nonsense, a violation of too many categories. Such opposite responses appear to confirm the possibility that we can (just barely) subjectively feel a qualitative difference in the cognitive style of each hemisphere.

The Power of Metaphor

The left's preference for linear logic is suited to immediate defensive or aggressive actions. But the context is a deeper well—one that is practically as large as the cosmos. The more context of which we are aware, and the more of it we use, the more accurately we can expect to devise survival strategies. This distinction is evident in the frequency with which the right's metaphoric insight registers deeper comprehension than the literalisms of the left; perhaps context knowledge is indeed one way of approaching truth.

Newton's explanation of celestial mechanics provides a classic example. He recognized that masses must attract other masses towards their centers, a process he called gravitational attraction. He was able to show that the force that pulled a ripening apple from a tree, on one scale, could also, on a vastly greater scale, hold a moon or planet in orbit. In this scheme, the familiar apple served as a metaphorical stand-in for moons and planets—objects that he could not grasp and bite. The apple thus provided valid mental leverage.

But Newton thought that action at a distance was an invocation of a mythical metaphor, a somewhat specious intellectual crutch; he did not believe that invisible forces could act across empty space. Unable to see further, however, he knowingly left the full explanation to later science. It was Albert Einstein who later treated gravity as resulting from a 'rubbery texture' of space-time, such that it became distorted into a bowl configuration by the central mass of a star.

As a result, a star's planets, can be considered to travel in straight Newtonian lines, but along the locally curved space-time. From outside the distortion, we see the effect as curved orbital motion.

Richard Feynman's use of a glass of ice water and an O-ring to explain a shuttle disaster was also metaphoric insight, and here it is interesting to reflect that other members of the investigative panel were prepared to rationalize away the claim that the Challenger disaster could be so simply explained. Feynman's visual show-and-tell analogy closed the deal. Had Feynman not been present, literalisms, i.e., rationalizations that supported hidden biases, might have ruled the day, as they so often do in committee deliberations.

How does Metaphor become Verbal?

When language evolved, some synergy between right and left hemispheres allowed metaphorical knowledge to participate in sentence structures. How did that happen? Stephen Pinker, in *The Stuff of Thought* (2007), suggests that all words actually began as metaphorical allusions—presumably in the silent hemisphere. The thought may therefore seem paradoxical because metaphors are eventually arranged into sentences by the left hemisphere.

Pinker implicitly explains the paradox by suggesting that, with familiarity, metaphorical terms become so closely associated with the circumstances in which they are habitually used that they shrink to what we term 'literal' words, losing metaphorical overtones in the process. In the shrunken state, they are tightly defined symbols that the left hemisphere can easily use in sentence structure.

Pinker's perspective leaves me with an impression of our vocabulary as a coral reef, with its bulk assembled from the literalized casings of bygone metaphors. The living, growing surface of the reef, however, still flourishes with the bright colors, the richer truths of the metaphors understood by the right hemisphere.

Metaphorical Connotations

In art forms that deploy metaphor, it is not essential for the listener to learn the poet's original source of context. As the many literally incomprehensible lyrics in Bob Dylan's repertoire illustrate, there is merit in triggering a personal search for connotations in the listener.

In these cases, we generally drum up some aspect of our own lives that fits within the metaphoric range, generating our personal Mr. Tambourine Man in jingle jangle time.

In other circumstances, however, it is necessary to understand more precisely the intended meaning of a metaphor, and to do that we must place it in the correct context. After all, many words have multiple meanings, both connotative and denotative, and so there is often a need to correctly ascertain which connotation is relevant in any given metaphorical phrase. Co-location of metaphorical skill in the hemisphere that also does broad context analysis seems to be a pragmatic distribution of neural expertise.

The Effect of Civilization

As man's cultures have evolved, the context from which most of us draw meaning has been changing from the natural world to a rationalized world of human artifice. A grid of city streets reduces the pressure on individual spatial skills. Social conventions and lists of ethical exhortations, such as the Ten Commandments, may reduce reliance upon the silent hemisphere's world-model.

But leaders may have become more manipulative and their followers less critically thoughtful. As a result, I suspect that there are now some among us who actively avoid input from their own right hemispheres. This suspicion originates from listening to those typically defensive individuals who seem to feel hunkered down in a mental pillbox, through the slit of which they spray talking points and one-liner slogans. The most striking part of this behavior is, to me, is the manner in which they then give out an impression of immediate status gratification, of seeming to believe that their half-witted ejaculations constitute appropriate contributions to the search for truth. The relevant thought, *Am I speaking bullshit?* ought to arise but it does not.

To the extent that the theory of democracy includes an assumption that every citizen can think clearly, we have a problem. An excessively left hemispheric mental style produces one-liner soundbites that avoid the sweaty tick-tock of actually thinking through and correcting our self-centered preferences in the clear sunshine of the bigger picture.

Metaphors of Broader Awareness

During the three score years and ten that we can expect to captain a corporeal body, it seems to behoove us to pay almost constant attention to the nuts and bolts of survival. Nevertheless, our most memorable moments occur when the broader perspective of The Muse is dominant.

Coming home late on a starry night, I usually stop to gaze out and think: my every constituent particle came from out there; I am already a spaceman. For teasing moments, I seem to project my thoughts back to the time when my atoms gleamed in a star's eye. But it never lasts long enough to truly satisfy, and I go indoors and climb in bed. In the morning, passing a spoonful of Mr. Kellogg's finest into my mouth, an incredulous thought may again leak in: I am holding a tool whose every atom was forged in the heavens, merely reworked in a factory, here on earth.

For a moment I am a God, with a virtual Excalibur in hand. I suspect that these fleeting moments of wonder reflect a brief pass through the right hemisphere—a momentary lowering of its defenses that allowed a glimpse of right hemispheric consciousness. My left side seems to resent the intrusion, and pulls me back. A mortal self again, I worry about literal milk dribbling into my literal beard.

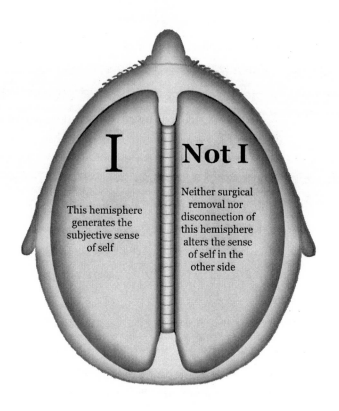

I | Not I

This hemisphere generates the subjective sense of self

Neither surgical removal nor disconnection of this hemisphere alters the sense of self in the other side

Chapter 25

The Muse and I

Naming the 'Not I'

In the interpretation we are developing, the left hemisphere emphasizes the differences between object-details, and at the same time becomes one: a self. The self then witnesses cause-and-effect sequences happening between objects, and produces sentence statements featuring 'I' as the chief actor and star of a show called *My Life*. In contrast, the right hemisphere uses a pattern analyzing approach that can better recognize commonalities across categories, and it thereby produces deeper metaphoric understanding.

There is a characteristic egoism associated with the self-hemisphere, an emotional concern that is usually called fear but which might also be construed as empathy for the self. In contrast, the right's broader concerns are associated with empathy for others, even fictional characters when they are imagined in the trials and tribulations of daily life. Although we routinely experience the ego's internal perspective, we rarely experience that level of intimacy with the right side's broader ruminations. Nevertheless, the general notion that the brain houses a dispassionate observer, whose broader perspective we switch to intermittently, has been raised many times in the past.

Adam Smith called it the *Impartial Spectator*. He felt that it lacked a self and therefore lacked a bias, allowing it to judge situations dispassionately—these were striking insights for 1776!

The Christian Brothers of my school days described a *Guardian Angel* that I now attribute to context analysis. For Carl Jung, it was the *Greater Self*. I like the term *Silent Witness*.

Still, the variety of these names is a clue to the creativity and holistic nature of this strangely removable and yet surely important half of our brains. For the rest of this chapter, I'll use the idea of the evanescent muse

Our most remarkable artists tend to be lonely, passionate, manic-depressive, and longing for the return of their muse. Many have stated explicitly that they do not feel that the self is responsible for their artistic success, and they reject any entitlement to pride in their work. William Blake felt that he could praise his own output because he thought 'William Blake' was no more than a secretary for 'Authors in Eternity.' Similarly, but with a different spin, Yeats had an elaborate myth that the creatures of Faery, the spirit world, have hidden knowledge but are unable to introduce it into the material world unless the poet offers himself as the medium.

Other poets have imagined that they become possessed by a daemon entity, whom they consider to be the real source of their most inspired writing. A similar notion is common among classical musicians (Mozart, Beethoven). Bob Dylan has described himself as simply 'taking down' songs that were already in the air.

How does this feeling of a non-self source of knowledge come about? Given the previous discussion, the question is rhetorical. The right hemisphere is a big sheet of non-self, with a known affinity with the metaphoric resonances of song and poetry, and an ability to add pleasing prosody. The right hemisphere seems to be The Muse's home, and the sympathetic vantage point from which artist's prefer to experience life.

Lorca: The Wandering Poet

Edward Hirsch, in his book *The Demon and the Angel* (2002), described the Spanish poet Lorca's realization that thoughts riding metaphoric horses transfer most easily between speakers who are on friendly, sympathetic terms with their audiences. Before his performance readings, Lorca tried to create these conditions with a ritual intended to appeal to what he called his duende. My dictionary translates the term as an allusion to a mischievous pixie. For Lorca, it seemed to have overtones of an accomplice in mortal struggles, like bullfighting. I don't know the nature of the ritual but when Lorca succeeded in the invocation, his performances benefited from establishment of a shared mental context, within which connotations were efficiently recognized. Hirsch makes a contrast with a flamenco singer, Juan Talega, who was an admirer of Lorca, but who was a more serene, orthodox, and emotionally restrained interpreter of his own art form. Perhaps limited by his personality, Talega only experienced the duende effect twice in his life. Both times they had to carry him out. When it happens, it is clearly powerful.

The Muse Takes Possession

In about 1984, I attended a Storytelling Workshop at the Augusta Heritage Festival in West Virginia. On the Monday morning, another participant forcefully challenged my notion that folk ballads were tightly crafted stories. She seemed to have had a one-sided mastectomy and wore a tee shirt emblazoned with the words: *Witches do Good Things*. Her manner was as relentless as Pete's had been in that long ago boxing match. I backed off and shut up, but resolved to demonstrate my meaning.

When it came my turn to tell a tale, I recited the lines of *The Band Played Waltzing Matilda,* written by Eric Bogle. It evokes the lethal stupidity of the First World War. I was conscious that it stirred within me something of the mood evoked when the older members of my family spoke of the Second World War. Speaking the lines of the initial verse went as it did in song, for the disconnection from self was being invoked but was not yet complete. During the next verse, however, the small hairs rose on the back of my neck. The poem is long and brutal, and alarming levels of cynical bitterness entered my voice. Soon my lips began to quiver and then distort, while perfectly apt inflections seared the words. Tears threatened—appropriate tears of a veteran left legless, as he bitterly watched a Poppy Day Parade that 'celebrated' a pointless war. On the edge of bawling, I staunched the tears and grimly continued. Now my neck muscles joined the dance, then my face and shoulders writhed to previously unknown rhythms and so constrained my respiration to their needs. And then it was over. There was no acting involved.

The silence in the room was prolonged; the women's faces were wet; the men just looked ashen. What had happened? Someone had once told me that poetry is best driven from the right hemisphere; now I had direct experience.

In this mode, the right had gained control of all my motor responses. I was once again a puppet, not of an electric fish, but of some sort of summation of all the pain and futility of war, which had become crystallized in Bogle's imagined veteran and which had resonated with my own family's stories.

The visible and audible effects in my body and vocal cords had also entrained a highly emotional part of the audience's separate minds. Of course, I now think that it was their right hemispheres acting in sympathetic mode. The rest of that day, and for many days afterward, the experience left me feeling deeply euphoric, as though some emotional circuitry had appreciated my Elvis moment. You will recall that I interpreted Elvis Presley's effect on his audiences in similar manner. Now I come to think of it, Presley's childhood favorite, *Old Shep*, had a great deal of the pathos and melodrama often associated with the right hemisphere.

When Bob Kerrey, a wounded Vietnam veteran, won his campaign for a Senate seat in Colorado, he chose to sing the same song, *The Band Played Waltzing Matilda,* during his acceptance speech. The newspaper report of that dry fact was inadequate and so serves as an example of the little acknowledgment the silent hemisphere's contribution to life is generally given in a society dominated by effort expended in the self interest of the left hemisphere. The self tends to assume that its literal prose version of what happens contains all the relevant information. (Caveat: As noted earlier, advertisers make as much use of the right hemisphere's susceptibilities as they can. But they are obviously doing so in pursuit of the profit motive and so the left's motivations remain dominant. A similar nuance plagues Hollywood's film industry.) I am struck by Bob Kerrey's obvious idealism in moving from wounded Vietnam veteran to Senator. I can imagine his right hemisphere being motivated to make sure that the mess in Vietnam would never happen again.

But in our day, more ill-founded wars have happened: one in Iraq under a false claim that weapons of mass destruction were present, and an immediately successful but inappropriately expanded war in Afghanistan. Neither effort turned out as planned. (Unless one counts Osama Bin Laden's intent to drain America economically.)

I believe that ideals need to win, even in a market economy, but they cannot do so if we don't first engage in the discussion of how our minds work to subvert idealism, a discussion that seems to have become more pressing in the present Great Recession.

An Ignoramus Goes Skiing

Like Juan Talega, I have had another experience with duende. It happened a decade earlier while I was at the University of Iowa. My friend Bob Benno had invited me to join a group of graduate students on a skiing trip to Minnesota.

Liverpudlians of my day had no exposure to skiing, and so my torso had no prior knowledge of the associated motions. But I took a beginner's lesson on the bunny slope, where I learned how to snowplow and turn.

I was somewhat embarrassed to be the largest and oldest beginner, and this may have inhibited me from asking questions. I just assumed that the snowplow was a standard braking maneuver, and that one turned to avoid trees. What follows is a perfect example of how we humans can practice a prolonged and detrimental neglect of context understanding. After an hour of practice, I joined another friend, Charlie, at the ski lift.

A scientist named Benjamin Libet had been part of an interdisciplinary training program at the neuropsychiatric institute of my graduate school days. (Yes, this is part of the skiing story.) Ben had performed a classical experiment, which showed that conscious information processing takes several hundred milliseconds.

Message: Conscious appreciation is slower than reflexes like the defensive eye blink, and it is wise to avoid rapidly changing novel circumstances that do not allow time for conscious self-correction.

Let's see... Charlie and I were standing on the top of a ski slope. In blind ignorance, I set off on my first real ski run, pointing directly downhill. A couple of seconds after pushing off, the existential thrill turned into pure panic, for I was already traveling too fast for conscious deployment of any control over the skis. Charlie had watched protectively as I set off and was horrified that another son of the sod had failed to grasp an important point.

Hoping that I would spill in a hollow about half way down the slope, he set off to pick me up, anticipating how he would pummel the basics into my evidently resistant brain.

Although my fear-filled left hemisphere had turned to useless mush, the right side seemed to come to my aid in the same way that it does when I jump off a diving board. It reflexively deployed the major muscles of my body in such a way that I stayed upright, and so I emerged from said hollow now doing cartoon levels of incompetent speed. This scared the bejeesus out of everyone in my line of approach to the ski lodge, which was unaccountably positioned broadside to the bottom of this slope.

As the grade leveled out, it was clear that ordinary friction over the available distance would not save me, but there were still hundreds of milliseconds before impact.

While contemplating the X of my final silhouette, my inner dialogue squeezed out: *Snowplow or die!* I forced my legs to splay, harder and harder, then leaned back. I came to a knock-kneed but gentle rest against the side of the building. I would never again ski so fast and so far and remain upright.

Completely dependent on my Silent Muse, which had driven my body flawlessly, I survived a half-witted approach to downhill skiing.

Charlie arrived and, of course, asked why I hadn't turned. "There were no trees," I said, slowly, as the light began to dawn.

The exhilaration that this experience delivered, with the sense of defying death, is the only thing I can compare to being in the grip of the daemon while reciting a poetic ballad. Lorca considered access to the duende to be integral to the survival of a matador when he challenges a bull with a rag and a rapier. I think I understand.

The skiing incident may illustrate a degree of motor control more usually associated with dance. So we have associated diving, singing, context revision, poetic metaphor, and now dance as specific animants of the right hemisphere. There is a complicating nuance: the left hemisphere can also drive its imitative version of all these things. But it lends them a self-conscious, and inauthentic cast.

The Hemispheric Tick-Tock

In the Shakespearian sense of *All the world's a stage*, we have today defined literal borders to rationality. Behind the curtains at stage left is the incomprehensible singularity at the start of the Big Bang. Behind similar curtains at stage right, we have similarly incomprehensible black holes.

Between these phenomena are sunlit boards upon which oscillating hemispheric crosschecks can keep track of a double cause-and-effect drama, our lives in their contexts.

Each cerebral hemisphere explores random resonances at different scales, for the purposes of self-correction and of teasing out new and thus creative perceptions. By crosschecking, each limits the other's potential for delusion. Thus the hemispheric tick-tock might be seen as the primary tactic of a creative rationality engine.

In this model of mentation, the timing with which a hemisphere is allowed to run free before performing a crosscheck emerges as a likely place for physiological and personality variations.

This is consistent with Pettigrew's discovery that the switching rates vary over a forty-fold range.

The Breadth of Rational Comprehension

There is an implication of the crosscheck conception that came to me late and we can fit it here. When one piece of a jigsaw puzzle is found to reciprocate the curves and colors of another, we lock the two together; and seem to enjoy our involvement in thus establishing a local zone of rationality. When someone recognizes a fit between the behavior of an apple falling from a tree and the behavior of the moon circling a planet, the zone of rational comprehension that then comes into focus is rather more substantial. To be more specific: a host of intermediate scale phenomena can now be fitted into the same scheme; they seem to come along for the ride.

It seems reasonable to expect that evolution would have discovered that increasingly broadened comprehension expanded the cognitive space for rational survival. The effect may have driven an increasing fascination with increasingly large-scale phenomena by the rght hemisphere and the complementary enthusiasm in the other side for dissecting finer and finer details. (Which I took to the electron microscopic extreme in my laboratory examinations of synapses.) When mutually harmonious explanations apply at the two extremes, there is an implication that everything in between is likely to be explicable along the same lines—a potentially huge zone within which one might practice survival-enhancing rational creativity.

The dilemma of the mismatch between the probability basis of quantum mechanics and Einstein's conceptions of cause-and-effect relativity illustrates the current limitation of human insight; the large scale seems inconsistent with the sub-atomic; and so we deduce that we must be missing something.

Purpose

Using an assumption that a cause precedes every event, we live our daily lives in a zone that seems to make coherent sense. Naturally enough, this has caused philosophers past to seek the ultimate cause. But note internal contradiction: the quest negates the original conception that everything has a cause. Meanwhile, science keeps pushing at the boundaries of large and small-scale knowledge—and both keep retreating.

As a result, we can now survey a spatial dimension of nearly fourteen billion light years, and the corresponding time scale—the age of the cosmos. In all the events laid out across this vast expanse of space-time, we see little evidence of progress towards an apparent goal. Instead we see events that are readily explicable in terms of random trial and error. For some commentators this has led to the conclusion that human affairs merely ricochet in a Sysiphusian pinball machine, with no net purpose to our existence.

I think Alice would have had little trouble turning this miserable viewpoint on its head. The collective evidence *is* consistent with a 'purpose' of discovering what is possible by trial and error. The stars the planets and life were possible, and have come to be. Moreover, the probability that random collisions of atomic material would create '57 Buicks is not zero but it is vanishingly small; fourteen billion years is nowhere near long enough to expect just one Buick to materialize on its own. And yet the human brain is acting as an accelerant of the cosmos-wide purpose of discovery, and Buicks came to be. Our brains, still practicing trial and error but with more deliberate selection of the options, can further accelerate the discovery of what has only been implicit in the past. With our executive capacity, we can catalyze external events just as smoothly as, on another level, muscle proteins catalyze the movements of our skeletons.

Our small-minded executive side prefers finite and monetarily profitable purposes laid out for achievement by a date certain. When it can't see one (beyond the frightened imperative to stay alive), it loudly complains. Perhaps we need to appreciate the scale of thought that is possible in the quiet side of the brain, and quit with the '57 Buicks and toxic fuels, in favor of deliberately contributing to the discovery purpose of the cosmos. This would surely be more rewarding than globicide.

Hemispheric Emotion

In the years before we recognized its study of context, there had been a tendency to map depressive emotions onto the right side of the brain. For example, it is a clinical fact that right-sided stroke victims frequently play word jokes with attending physicians. This has been described as 'euphoria' and that seemed consistent with the idea that the left verbal hemisphere normally drives positive emotion. From this perspective, it was believed that right-sided stroke patients failed to display the expected depression because the 'ability to be depressed' had been correspondingly wiped out by the damage. All the emotion that was left was the residual and rigidly positive mood of the verbal hemisphere.

The latter analysis ignores the silliness of the supposedly logical hemisphere being in a good mood after a right hemispheric stroke. Furthermore, the frequency with which puns elicit groans undermines the claim that they connote euphoria.

Laughter

Before I suggest a more likely division of hemispheric emotions, it will be useful to consider laughter explicitly. While generally producing complementary perspectives, it seems inevitable that hemispheric interpretations would occasionally conflict.

A David Brown cartoon in *Playboy* years ago provides a good example. A filling station attendant is inserting the fuel pump hose into an opened flap in the rear leg of a huge dinosaur. The creature's elongated neck is curled back as it looks on approvingly. A station sign that arches over the scene says, "FOSSIL FUELS."

While the scenario evokes curiosity, humor is only felt when the words reach consciousness. What has happened at that point? My guess is that the intuitive hemisphere takes in the context of the whole scene and sees a whimsical difference with its world model, for example, the fact that dinosaurs and gas stations didn't overlap in time. This may be entertaining but it isn't particularly amusing. But when the left hemisphere searches the short chains of logic in the scene, like the words on the sign, that logical tissue seems to find rationality: a fossil has come to be fueled, as the sign says, moreover a fuel hose is being deployed in the recognizable fashion; *What's the problem?* Laughter arises at that point. It is as though the context-aware hemisphere recognizes that the linear thought snippets of the verbal side are too narrow. (The fact that jokes are eviscerated by literal verbal dissection also implies the presence of a fast, intuitive, nonlinguistic, and usually more adequate interpretation.)

In this view, laughter labels a particular class of hemispheric crosschecks. What is so special about them that they are singled out with a fit of giggles? I suspect that laughter signals that the left's loose associations are sharply violating broader context wisdom. In other words, I suspect that, in the face of apparent left hemispheric oversimplification, the normally silent right hemisphere breaks it own taboos and delivers an overt disparaging vocalization—laughter.

The chuckles seem directed at the self and seem to ensure that the right's negative insight is recognized, whether the left was inclined to ask for a crosscheck or not.

This interpretation goes a long way towards explaining hostility to laughter on the part of self-righteous demagogues, those emperors with skimpy wardrobes. Laughter erupting from underlings has a dangerous potential to catalyze widespread crosschecks and sudden eruption of broader rationality, with consequent weakening of demagogic control.

Some split-brain patients were also reported to giggle in response to images (such as nudes) presented selectively to their right hemispheres. They could not orally explain why they were amused because their left hemispheres had not seen the same images, so here we have another example of (somewhat embarrassed) humor felt and expressed by the right hemisphere.

On the basis of a claim that we don't laugh when we are alone, a current psychological study of laughter has emphasized its social function. The theorists consider that laughing while reading a book is a social event, which seems to me to be an unacceptable stretch. But if the two cerebral hemispheres are regarded as the first 'society' in which laughter found a function, then this 'social' view fits the above construction.

A Sticky Switch?—Nojokia

Individuals whose psychology does not allow them to "get" jokes are fairly common. Their normal social behavior is unimpaired, but they seem to check their context assumptions slowly. They are startled when those around them erupt into laughter, and they typically beseech a verbal explanation, which only continues an emphasis on the verbal hemisphere and is therefore unfunny. Has the left's emphasis on self-control overridden access to the corpus callosum? Has the right short-circuited in some other fashion? Are these the individuals in whom Pettigrew has measured the slowest hemispheric switch times? I don't know, but these are good questions.

Hemispheric Emotion Revisited

Now let's return to the right-sided stroke victim half-paralyzed and yet 'euphorically' punning away in his hospital bed.

I see punning as a minimalist form of humor and lean toward the notion that an intact and readily accessed right hemisphere is necessary for the belly laugh merriment that arises when a profound switch in perspective is triggered. In this view, Victor Borge's approach to classical music with the slapstick sensitivities of Charlie Chaplin beats a pun any day.

Furthermore, notice that each of the conflicting pun interpretations is verbal, so both elements of the conflict are taking place within the talking hemisphere. The simplest interpretation, and therefore the one Occam might prefer is that such stroke patients are in the grip of a verbal hemisphere that can no longer participate in a tick-tock with a context appreciating and crosschecking partner system.

On the 'Depressive' Right Hemisphere

The opposite syndrome, seen in a left-hemispheric stroke victim, is depression shown by crying and swearing. This has also been interpreted as evidence that the right hemisphere 'drives' depression! A more likely interpretation is that the context analyzer has correctly appreciated the context and is simply expressing the appropriate emotion.

For bipolar patients, however, prolonged episodes of being in right hemispheric mode definitely correlate with the sensation of depression. In these clinical cases, it seems appropriate to consider that the right hemisphere itself is depressed. But this is hardly evidence that the physiological character of the right hemisphere in all of us is intrinsically depressive. As we have seen, Jill Bolte Taylor spent a long time in there and was anything but depressed.

An Alternate Cartography

The inner dialogue, which is associated with emotions like fear, anger, greed, envy, arrogance, and pride, invariably stresses the 'I.' This emphasis implies that these emotions will eventually prove to be driven from the left hemisphere. In contrast, the dominant emotion widely recognized to be associated with the right hemisphere is empathy, often expressed as resonant identification with the afflicted, often to the point of crying together. To the extent that there is any associated dialogue, it is either couched in expressions of sympathy for, and offers to assist the 'other.' or, as noted earlier, in the form of empathetic poetry.

Nirvana, which may be construed as empathetic identification with the cosmos (c.f. Jill Bolte Taylor) can also be associated with right hemispheric consciousness. This thought can be taken a step further. I will frame it as a syllogism, as follows.

If the left hemisphere's preoccupation with the organism results in its identification with the self, what might follow from the right hemisphere's study of external reality?

In Viktor Frankl's *Man's Search for Meaning,* 1959, he converts his horrendous concentration experience into a philosophical nugget: *It does not matter what we expected from life but rather what life expected from us.* I am inviting a similar conclusion from a far more benign exposure to experimental neuroscience. In my case the thought arises in connection with my response to the above question: Our right hemisphere may have a capacity to identify as strongly with the cosmos as our verbal hemisphere does with our self-consciousness— and this may underlie the cosmos-consciousness reported by Jill Bolte Taylor.

When I began writing I had no suspicion that exploration of the bullshit problem would lead me to this philosophical epiphany.

Chapter 26:

The Cross Check Criterion

In this, my closing chapter of *Headwaters*, I will attempt to fill a gap in my thesis with what will be its most speculative component. If the verbal hemisphere's question "Will this work?" is addressed to the silent hemisphere, how might that tissue be able to tell if an idea would 'work'?

Stephen Wolfram: *A New Kind of Science* (2002)

Science aims at a rational search for the simple patterns presumed to underlie reality. The Periodic Table is a good example. Dmitri Mendeleev, an early Russian chemist, used playing-card stock to write down the characteristics of the known atomic elements. When he played a game of chemical solitaire with those cards, he came up with what we now call The Periodic Table of the Elements. Later scientists discovered the underlying rule: each successive element has one more proton and one more electron than its predecessor (along with a variable number of neutrons). It could hardly be simpler, and yet the complexity of chemistry, including the chemistry of life, flows from that foundation.

The success of reductionist thinking in physics and in chemistry suggested to Stephen Wolfram that reality might display a limited number of patterns arising from simple underlying rules. Making a huge counterintuitive leap, he wondered if one might ignore the generally complicated epiphenomena that develop, and concentrate instead upon underlying potential simplicities.

Wolfram's laboratory technique is not unlike that of Jennings, who watched protozoan behavior with his eyes glued to a microscope. Wolfram watches how rules affect tiling patterns on a computer screen. He has investigated all the simple rules imaginable in the hopes that the "universal rule" will turn up (and be recognized when it does). All this dazzled me; it seemed simultaneously brilliant and simple. But I was pretty sure that it was above my pay grade.

How has Wolfram Been Doing?

Wolfram's automatons are allowed one somewhat Shakespearian action: to shade or not to shade a grid box within a matrix. The simplest rules produce tiling patterns suitable for the floors of public toilets: all black, all white, or checkerboard alternation.

This prosaic beginning prompts one to suppose that more complex patterns will require complexity in the rules themselves. Yet further experimental results quickly confirmed scientific history: there are simple rules, even in this simple circumstance, which can generate apparently endless, seemingly random complexity.

Most interestingly, some of these simple rules produce self-referential patterns similar to the fractal patterns discovered as an outgrowth of a branch of mathematics called 'Mandelbrot Sets.' These are as variably complex as snowflakes and branching vegetation.

In *A New Kind of Science* (2002), Wolfram shows that many biological processes appear to follow rules akin to computer automaton rules, producing closely similar patterns. Although the constraints that force conformity (to the rules) are electronic for the computer automatons, and biochemical in the case of living organisms, the similarity of the output provides eerie support for the view that there may be a limited number of rules and patterns in our context, that is to say, in our four-dimensional reality.

What is the Fundamental Form of the Crosscheck?

Let's look back a moment. In the 1960s, personality extremes were often thought to be expressions of the cerebral dichotomy.

Obsessive logicians, fortuitously well adapted to the lifestyles that accompanied Industrial Revolution, were said to be dominated by the left hemisphere. On the other hand, there were 'irresponsible' artists starving in their garrets; they were said to be laboring under the dominance of their right hemispheres. In this widely gossiped view, there was no direct implication that one side of the brain had evolved to correct for the errors of the other. The palpably improved *text* versus *context* dichotomy does suggest a continual reference of one side to the other, but it still fails to illuminate the underlying 'mechanics'. Of what might the actual technology and philosophy of a hemispheric crosscheck consist?

If reality is inherently full of self-referential, rule-based patterns, as Wolfram suggests, then the brain may have long ago succeeded in recognizing and using them. A peculiarity of fractals is their ability to vary in detail at different magnifications, while remaining resonant with their parent algorithm. If fractal comprehension was an early form of brain language, and if fractals are capable of portraying potential executive actions, then the left hemisphere might be able to preview or check executive possibilities against a 'parent' fractal developed from broad comprehension of external reality. The latter would be expected to reside in the right hemisphere. *If so, resonant harmony between such pairs might be equivalent to feasibility testing.* Yes, there were a lot of 'ifs' in the above account, and it was hard to state that thought in words, so I turned to the graphic on the following page. It presents the suggestion in a perhaps more plausible fashion.

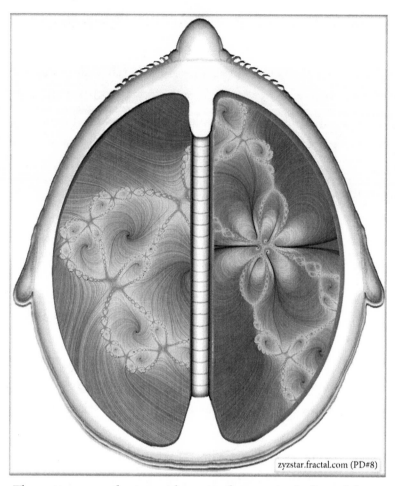

The patterns on the two sides arise from a single fractal form—at a higher magnification on the left. I suggest that micropatterns of accurately comprehended details on the left, the home of the executive self, when processed by some unknown but built-in algorithm, should always resonate with the corresponding macropattern of accurately understood context knowledge on the silent right side.

This suggestion implies that the full version of the humble query, *Will this work?* is as follows: *Are the fractal details of my new conscious idea in resonant harmony with the big picture fractal that describes the world—that you in the right hemisphere have studied more than I in the left?*

322

Of course, it is very unlikely that my example of a randomly chosen fractal, from the public domain, has any great connection to reality. On that score, we will need to wait for more insights from people like Wolfram. But for now, we have made the idea of hemispheric crosschecks a bit less vague, a bit more meaningful.

Late News: The IBM Computer Watson

IBM has recently reported a relevant experience while developing the computer Watson. Watson was designed to have access to virtually all the library-based loose associations available to our civilization, and to be able to use them in the context of *Jeopardy,* a television quiz game. In its first trials, Watson rapidly came up with stupid loose correlations, and lost the game to its human competitors. I doubt that it was smart enough to feel humiliated; nevertheless, it was sent back to the woodshed. The IBM team then developed a supplementary suite of software that enabled Watson to determine the probabilities of correctness that should be associated with each of the initial associations. Reprogrammed to use the most probable it now outdoes human competitors.

This pragmatic evidence of a need for a division of computational labor fits our general thesis well, The supplementary system in Watson's case seems to calculate the number of cross references supporting each of the original loose associations. This sounds somewhat akin to the notion of large and small–scale resonant match. How far does the supplementary software go in modeling the way the world works? Can it do film stars in evening gowns? Could such a device become the cornerstone of a great institute of rationality, a massive context conception for general reference?

The Social Animal (2011)

In David Brooks' book, *The Social Animal* (2011), recent insights in cognitive psychology are illustrated in the life story of a fictional couple. In his Chapter 13: *Limerence*, Brooks describes the mental events happening during the couple's courtship. 'Limerence' is a newly coined term for 'a sense of harmony driven by squirts of dopamine onto pleasure neurons in the brain.

With this concept, Brooks characterizes the brain's struggle to find models of the world that align with the daily stream of passing experiences. One of these activities is the anticipation of rewards. When they arrive as expected, a 'kaching' of harmonious resolution, limerence, arrives in various brain nuclei. Brooks suggests that the premonitory expectations themselves, the earlier daydreams, deliver small squirts and, presumably, anticipatory limerence.

I can recast Brooks' description in the terms of the crosscheck that is described in the present chapter. When a mental model of the future imagined in the right hemisphere are matched by actual sensations from current experience, resonant harmony between the hemispheres amplifies dopamine release and so generates the subjective sensation of limerent gratification. Brooks ascribes much of our motivation in life to this search for limerence.

Despite considerable affinity, in the final analysis, Brooks and Boyne arrive at different conclusions. Our diverse notions arise from contrasting views of the merits of conservatism, a fuller exposition of which is planned for my follow-up text. (I should here note my theft of Brooks' term 'status sonar' for that pervasive left hemispheric preoccupation.)

Globicide as Behavioral Pathology

Before we biochemical creatures began to haphazardly distribute previously unknown chemicals into the biosphere, or before we released long-fossilized carbon back into the atmosphere, it would have been preferable to work out the quite predictable deleterious long-term consequences that we are now learning to understand the hard way. Instead, our forebears emphasized immediate gratification and blindly opened a Pandorian Box. I attribute their failure to act with appropriate forethought (and our continuance of the tradition) to the lack of a simple instruction manual for the brain.

In lecturing on behavioral pharmacology, I often invited my audience to contemplate the behavior of a person sitting next to them. I asked whether each listener thought that any peculiarities present justified therapeutic intervention, and so initiated a general discussion of what constitutes the range of normal behavior.

The professionally pragmatic conclusion is that behavior with serious negative impacts upon one's social or work relationships is a candidate for either psychotherapy or drug treatment. Anything less might be idiosynchratic but might also be preferable to conventional behavior, so should not be 'treated'.

On these grounds, epilepsy is considered to be a behavioral problem deserving of drug treatment. Another example: I tend to faint immediately after receiving an injection, a nuisance behavior which my dentist eliminates by ensuring that I have taken a single dose of the anti-anxiety drug valium thirty minutes earlier. Psychiatry's Diagnostics and Statistics Manual is full of further categorizations of troublesome behaviors and corresponding treatments.

Now for an awkward question. If a technologically competent species arises on a planet and proceeds to use technology for short-term gratification—with a net consequence of planetary degradation and potential globicide—is not this the mother of all behavioral pathologies?

And if it is, does it indicate a need for a species-level of therapeutic intervention?

The question is surely unwelcome but just as surely relevant. Moreover, the answer is: *Yes, the threshold for therapeutic intervention was reached long ago.* Is there then a potential drug treatment? If we confidently understood the mental limitation that makes us vulnerable to short-term rationalization and world wars, then pharmacotherapy might be conceivable. The thesis I am offering is certainly of this kind: it suggests that too many of us are overusing the verbal hemisphere and so becoming half-witted devotees of its one-liner logic. Either a drug that inhibited left hemispheric self-righteousness, or one that activated right hemispheric restraints might be a candidate therapy.

Is this outrageous? Widespread caffeinated beverages may already contribute to the overstimulation of the executive side of the brain. In that case, therapeutic intervention might take the form of recommending decaffeinated substitutes, so no, this manner of thinking is not at all outrageous, it is merely unfamiliar. My purpose in writing this book, however, is to suggest a far more invasive psychosurgical procedure: enlightenment based upon deeper comprehension of the brain; in other words, truth-as-logotherapy as Viktor Frankl might express it. We are not creatures plagued by a Devil nor are we favored by an evident God, we are merely vulnerable to rationalizations along these lines. The plain fact is that we are the world's dominant species with an unfortunate evolutionary heritage that encourages us to rationalize in the selfish interests of short-term dominance battles, even over our favorite creation myths.

The banner that I am trying to lift was first flown thousands of years ago, by the prophets of the Old Testament. Alarmed at the dominance of shortsighted greed and state-sponsored arrogance, they sought to promote empathy, especially for the poor, the downtrodden and the victims of hard times.

Those prophets may have instinctively recognized the need for a better balance between the two sides of the brain, but their evidently helpless ejaculations, such as Amos', *Behold, the word of the Lord!* were further examples of inadequate context understanding.

(That message may well have represented an impassioned imperative originating in Amos' right hemisphere, but I don't think it was the word of the Lord.)

I hope the now desperately needed anodyne can be found in modern accounts of the vulnerabilities of language and the brain. In this regard, Leonard Shlain's *The Alphabet versus the Goddess: The conflict between word and image* (1998) merits great credit. Shlain pinpoints the Ten Commandments as the moment when alphabetic script focused the left hemisphere's language specialization on scriptural religion. There was a remarkably sudden shift in the tone of spirituality towards male egotism. This included blaming woman's alliance with the devil for man's ejection from the Garden.

In *The Master and his Emissary; The Divided Brain and the Making of the Western World* (2010), psychiatrist and academic, Iain McGilchrist extends Shlain's thesis. Also a twenty-year project, this is a masterful review of the literature. McGilchrist also concludes that the left hemisphere has now become detrimentally dominant.

Neuroscience and Religion

The great religious denominations were seeded by individuals such as Confucius, Buddha, Christ and Mohammed, all of whom seem to have experienced a cosmos consciousness that delivered broad empathetic understanding of the human condition.

Lacking any knowledge of brain physiology, however, the mystics and their disciples tried to transmit their insights using devices such as parables and 'eight-fold-ways.' These answers to the problems of evil and man's self-destructive ways proved inadequate.

In every case, instead of promoting a rational spirituality, their followers fell back on self-serving rationalizations. The spiritual enlightenment bequeathed to the generations was thereby corrupted and dissipated. In what is called the scientific method, however, potential insights and explanations are subject to experimental tests. Results are presented in open forums that regularly cleanse science of personal dogmatic accretions. Over recent centuries, the resulting, ever-broadening comprehension of the astounding natural world, from the sub-microscopic realm to the vastness of the universe, has acquired spiritual overtones. For example, Hubble's images of deep space can be experienced as dogma-free cathedrals. It seems that both Elvis and spirituality have left the church building; neither seems likely to return to such cramped quarters.

There remains a problem. Even in a post-Darwin, post-Einstein age we have provoked a mismatched conflict between our corporeal selves and the ground of our being, the planet Earth. I began probing our situation by trying to trace the sorce of bullshit, gradually recognizing that inadequate consultation with the right hemisphere was the more fundamental cause of a general behavioral pathology.

I have been tempted to blame a 'design flaw' in the brain. But our brains are the product of billions of years of evolution's life-and-death research and development program; I have no better brain to offer. Have a few decades of human experience in bioengineering, as in Dolly the sheep, been adequate to support an engineered improvement of our current inheritance? I can place zero faith in efforts to redesign what evolution so patiently tested and developed. Consequently, I settled on the term 'design vulnerability.'

Beyond acknowledging vulnerability, our hopes for a rewarding future will require a description of our context and deliberate purpose, one that we might all agree to use as a reference base when crosschecking the rationality of our individual endeavors.

That description must be as basic as possible and open to continual updates. I hope it acknowledges the extraordinary spiritual potential of our right hemispheres. I can, of course, hear a catcall of derision swelling in a peanut gallery—*That's too simple!*

But consider this statement of our present implicit purpose:

"To inflame and then cater to consumptive human appetite, so that the rich get continually richer while the planet itself is plundered of its natural capital, and the poor can only hope to be trickled down upon."

If inflaming desire so as to enrich above and trickle below is no longer our intent, then that position needs explicit repudiation, after which something of matching concision and universal appeal needs to be fashioned, and then broadcast to all humanity.

Apart from deflecting further tragedy and escaping calamity, there is something here that is philosophically exciting. If and when we manage to close our alignment gap with our planetary context, using a brain with the evolutionary provisions already built in, we might also slip into register with the cosmos, perhaps experiencing as startling a transformative upgrade as that which attended the cry from Billy Batson, the crippled newsboy:

"Shazam!"

(And with no bioengineering involved.)

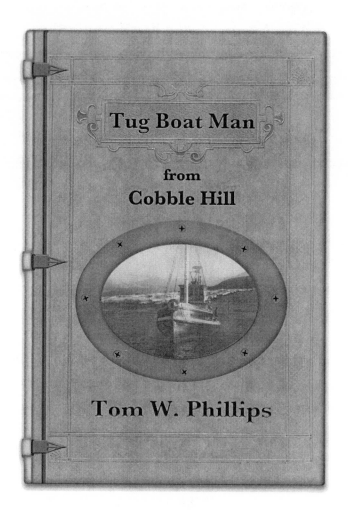

Chapter 27

The Tug Boat Man

An Unexpected Collaborator

I have had an unexpected collaborator through the years of writing. Like me, he was a World War Two baby, who quickly grew puzzled that things weren't turning out as well as they might.

Tom's soldier father came back from the war as an angry man who was determined to stay that way, and Tom's mother quickly left him, taking Tom with her to live in Montreal. Unfortunately, housing conditions were poor, and she came down with tuberculosis. In those days, TB patients were 'treated' with fresh air and kept away from where they might infect others. Tom, eight years old, watched the result back at his grandmother's cabin on Vancouver Island. A chicken coop outside was lined with linoleum and newspapers, and this was where Joan gradually weakened and died.

Raised by his grandmother and a step-grandfather, Tom grew up in poverty. Unlike myself, he was aware of being poor and that made him reluctant to make friends, for he anticipated dying of embarrassment if they came to his home and wanted to use the bathroom. Now in retirement, and comfortable with his own company and his good wife, (Social Security works!) Tom spends his time thinking, drawing, writing and making music.

Hearing him sing at an open mike session over twenty years ago, I recognized his authenticity immediately. When I also learned that he had made a tape of his songs, I went begging for it. We met again in Ship Yard Cove, a local marina, but it was not until he retired that our friendship became close. There were then regular meetings at my house with a group of writers, and I invited Tom to come, imagining that he might share a poem or two. He was reluctant— the old shyness. Moreover, he knew that the rest of us had been to college, whereas he'd left school as soon as he could work.

Nevertheless, one day he came. He had a sheaf of paper in hand and we were treated to a monologue that would have done credit to Mark Twain. Of course, we entreated him to come back and he became a regular. With an evidently easy access to his right hemisphere, Tom expresses himself in the now gentle, now strong intonations that are informed by the life lived behind the words.

We have all responded to self-conscious performers with speedy applause, intending to be encouraging. Paradoxically, this betrays a left hemispheric, status-aware state of mind, and a relative failure of the artistic intent. Tom's presentation naturally entrains, even hypnotizes the audience's right mind. In an involuntary tribute to his final thought, the typical response is a stunned silence, an effect I've been happy to witness over and over again.

As years went by, with no sign of the well running dry, I learned that Tom had initially brought stories from decades gone previous, still unpublished because editors couldn't handle his unique punctuation. Having found a receptive audience, he had now been re-bitten by the bug and was writing fresh material. As he became my writing, talking and singing example of hemispheric switching, I began to have a niggling feeling that I ought to arrange for everyone to enjoy this.

That niggle might well have gone the way of most niggles, but for something else. Tom has also worked at getting to the bottom of the basic question of why so many things don't turn out well. We thereby form an odd couple: a poet-in-the-street raising and re-raising the question to which a scientist-in-the-street has been trying and retrying to develop an answer.

When I just couldn't stand it any more, I asked Tom if I could record him. He agreed and we had a grand time making several CDs, with an audio book memoir still in the can. Book production technology has reached the point where the punctuation, language and illustration preferences of the traditional publishing houses can no longer keep one light's hidden beneath a bush. I've assembled some thirty of Tom's pieces into a brazen book: *Tug Boat Man*. There will be an on-demand-print form, a Kindle version and a multimedia ebook extravaganza from the ibookstore. This includes an audio track of Tom reading his own words in his inimitable way.

Chapter 28
Goose Brains

by

Thomas W. Phillips

It is five a.m. Ships are moving across my window, miles distant. Slowly they weave their wakes up and down Juan de Fuca, Rosario and Haro Straits. From my vantage point I can see the confluence of the waters beneath them making their way to the Pacific Ocean by way of Cape Flattery on the outgoing tide.

Some ships round Discovery Island southeast of Victoria and continue up Haro Strait to Vancouver, while others turn south into Puget Sound, or east to Rosario—bound for Bellingham or the refinery at Cherry Point in Padilla Bay.

Tankers, container ships, cruise ships, military vessels and private craft of all sizes and descriptions ply these channels twenty-four seven. I wonder: *Where have they come from; where will they go when they leave?*

A flight of geese passes overhead in their standard formation. The northern migration calls them to the lakes and rivers of Canada and Alaska.

Like the passing ships, the geese have ports of call on their journey, places to rest and refuel. "A long skein of beating wings, with their heads towards the sea," the old Scot's song says—toward the great northern sea at the top of the world that nourishes everything in a grand expanse at which we fragile, frightened souls can only marvel.

How do they respond to 'time' and 'distance'? There must be a schedule, a program. Surely a journey of such magnitude cannot be left to chance. Which of them assumes command? I imagine it is the most experienced, the steadiest and most levelheaded of all, with a keen sense of purpose. Is this intrepid leader self-appointed, or chosen by others who recognize his capabilities and depend on him entirely to keep them from danger and dire consequences? A puzzle, to be sure, but successful these thousands of years past. An example to follow; something to ponder.

"Silly goose," they say. I wonder how silly. One always stands guard while the others feed or rest. Are they silly who lead and are led without dispute or dissension on a recognizably correct course? Who travel the world in peace, with purpose and direction? Who are loyal and steadfast in their relations with one another, leaving others of their kind to go about the business of their lives unmolested? Wise and capable, they protect their young and are entirely devoted to their mates; something we find optional.

I once heard a man say, "You don't have the brains God gave a goose!" I guess that says it all... In our headlong rush to control and manipulate everything, we are missing the message. The visionaries look too far into the future to solve the problems of the present.

Next time you see a flight of geese, on their way to their appointed destination, think about it... *Where are we going?*

Corrections

Liverpool's May Blitz

Storytelling was the original form of entertainment and to stretch the truth in the service of a tale was entirely respectable. I eventually learned that Liverpool's May Blitz happened in 1941. Since I was born in January of 1943, I couldn't really have been conceived the night the mouse died. I suspect that Mother fuzzed up the chronology because she thought it made for a better story. Since I agreed, I didn't fix it.

'Binky' the 'Canary'

Reviewing audiotapes recently, I re-learned that the so-called canary that needed rescuing while bombs dropped was not a canary at all. I now suspect that the story as I first heard it was being muffled by the floorboards of my bedroom where I was supposed to be asleep. Hearing of an animal called 'Binky' I seem to have assumed it was a canary. (At least they rhyme.) And, of course, a bird would have needed a cage.

In fact, it was a dog owned by my grandmother's sister Aunty Tot. If he got out of the house during air raids, Binky was prone to run around Bootle barking at the bombs, which didn't help. So he was 'rescued' from open fields and taken into the shelter. Moreover, it was mother that did the rescue. Once again, I decided to leave it the way it had been in memory.

Fortunately, I didn't need to embellish science in quite the same way. The challenge in telling those research results is to make the unexpected and fantastic seem prosaic enough to be understood.

Acknowledgments

Plan B, demobilization from academe, gave personal sovereignty, as well as the total freedom from grant proposals that independent thought seemed to require. It would not have been possible, however, without the academic pension provisions built up during my active years, for these became an annuity that supported me while building and writing. I thank all my former colleagues at Northwestern Medical School for those delightful years in their company, particularly present Chairman Gene Silinsky, whose pivotal influence I described herein. Former graduate student, now Professor, Tom E. Phillips, at the University of Missouri, Columbia, I thank for the thesis collaboration that helped make Plan B viable, and for encouraging the present book.

I am indebted to the late Fred and Marilyn Ellis of Shaw Island, who heard me lecture on these issues twenty years ago at the U.W. Marine Labs; they also encouraged the notion of a book for the general reader. I wish I'd homebuilt faster, finished the writing sooner and that they had lived longer.

Thanks are due to many people for their willingness to provide substantial critical input. First prize goes to my editor Jim McCormick, without whose blunt patience I would never have gotten anywhere. Thanks also for a detailed critique from Liz Francis that led to a major revision. With regard to the thesis, Beth Helstein belongs in the usefully skeptical camp; I hope she finds the final version more creditable than the beginner's mess that she kindly plowed through.

Dennis Willows, former director of the University of Washington Marine Labs in Friday Harbor, another knight of the order of the microelectrode, gave encouragement and cautions when I most needed them; he wanted me to moderate my social and political criticisms in important ways, which I hope this volume does.

Hercules Morphopoulos provided the corrections noted to the tale of the dental school and arranged further welcome contacts, including Michael and Susan Lopez in San Francisco.

John P. Geyman, M.D., Professor Emeritus of Family Medicine, University of Washington and author of *The Corrosion of Medicine: Can the Profession Reclaim Its Moral Legacy?* gave a welcome endorsement.

Thanks to Steve Simpson, spiritual thinker and former director of the Port of Friday Harbor for several suggestions and encouragement. I was delighted to receive appreciative comments from my old chairman at Northwestern, Les Webster, and from a former colleague there, John Mieyal, both of whom are now at Case Western Reserve in Cleveland. John Disterhoft at Northwestern, although cautious, was at least impressed with the house building adventure. I thank another early retiree neuroscientist, Peter Snow, who is the author of *The Human Psyche in Love, War and Enlightenment* (2009), for extended conversations and useful ms criticisms.

Tony Vivenzio gave useful advice, and introductions. Three get special notice for absorbing the whole thing in successive days of reading: John Geyman, Jim Nollman, and Ernest Pugh. Victoria Foe almost qualified but she was on a plane bound for Australia and may be considered a captive audience. Since she personally witnessed some of the electron microscopy described and liked the book, however, she still counts.

Thanks also to Colin Hermans for a delightful dinner and discussions. Supportive comments, advice and corrections from the following fellow islanders were deeply appreciated: Paul Ahart, Nate Benedict, John and David Brown, Keith Busha, Tim Cowell, Barbara Cox, Kirsten Crawford, Terry Domico, Dan Finn, Peter Fromm, Ted Hope, Pam Herber, Dick and Nancy Hieronymus, Sue Hill (who found the melanoma in time); Kay Kohler, Rod Kuhlbach, the late John Heric, Joe Hudson, Roberta Leed, Samantha Leigh, Lyn Loring, Macro, Colin Megill, Ted Middleton, Mary Nash, Jonathan Piff, Emily Reed, Stephen Robins, Ron and Jean Shreve, Dale Sitts, Jim Slocum, Meg Strathman, Julia Thompson, Marilyn, Seth and

Wyeth Styles, Lee Sturdivant, Janet Thomas, Linda West, Dory Westhoven and Susan Wingate.

A particularly unusual piece of input deserves note. Jackie Altier, a member of our writer's group, was once a Hormel Girl. These were a chorus line of dancers and singers that toured the United States in the 1950s attempting to sell Spam. (I'm serious!) This opportunity led Jackie away from her Chicago hometown to Hollywood and the glamorous life of a singer. Upon hearing of my thesis for the first time, Jackie alerted me to E. Hirsch's text on performance, *The Demon and the Angel*. My response to that book was basically the entire genesis of Chapter 25: The Muse and I. Thank you, Jackie!

Farther afield, thanks to Jim Feeney and Doreen Preece (England), Graham Barry (Australia); Hanns Becker and Antje (Germany); Rosalind Becker (Northern California); Maureen Keller-Taylor and Pete Dingsdale, (Southern California); former student David Oldach (North Carolina); and Barbara Svoboda (Maryland).

My coffee shop and dinner party companions of the last twenty years, Bob and Lorna Dittmer, Ed Stiles, and the late Jim Hudson all contributed immensely to the fact that this book was ever written. (Lorna, in particular, helped me to take, punctuation. More Seriously.) None of all these people should be blamed for residual mistakes, weak inferences, and controversial opinions. Those can all be blamed on Dick Cheney.

Taylor Bruce got the text illustrations underway; her work is featured in the greyhounds of Chapter 2. Tom W. Phillips provided many more. Unsigned diagrams and the house illustrations are mine, (redrawn as noted if based on other sources). The fractal brain illustration is based upon public domain fractal #8 provided by kosoru.com. The drawings of children were absconded from Albert Hendschel, now public domain.

Similarly, the seal in Chapter 24, and the shovel-avoiding hens (Chapter 5) were extracted from *Jardine's Naturalist's Library* and are so old that retroactive claims of copyright cannot reach them.

The electron micrographs of Chapter 18 are from two research papers, one in *J. Neuroscience Methods* (1979): the other in *Journal of Electron Microscopic Techniques* (1984); permission to reproduce these has been requested. The underlying research was supported by an originally three year NIH grant, which I stretched like taffy between 1977 and 1983: NINCDS NS13043.

When the text was finally ready, the island provided another essential asset in the form of W. Bruce Conway of Illumina Publishing; thanks to Bruce for technical know-how and advice with the mechanics of book creation and design.

References

Ackerman, D. 2004. *An Alchemy of Mind: The Marvel and Mystery of the Brain.* Scribner.

Almli, C.R., and Stanley, F. 1987. *Neural Insult and critical period concepts.* In M.H. Bornstein, (Ed.) *Sensitive Period in Development: Interdisciplinary Perspectives* pp.123-43. Laurence Erlbaum Press.

Austin, G., Haward, W., and Rouhe, S. 1974. *A note on the problem of conscious man and cerebral disconnection by hemispherectomy. In Hemispheric Disconnection and Cerebral Function.* M. Kinsbourne and A. Smith, Eds. Charles C. Thomas Pub.

Bartlett, F.C. 1932. *Remembering: A Study in Experimental and Social Psychology.* London. Cambridge University Press.

Bogen, J.E. 2006. *Intermanual Conflict. History of Neuroscience in Autobiography 5.* Pp. 94-95. L. Squire, Ed. Elsevier Academic Press.

Bohan, T.P., Boyne, A.F., Guth, P.S., Narayanan, Y., and Williams, T.H. 1973 *Electron Dense Particle in cholinergic synaptic vesicles.* Nature (Lond) 224:32.

Bolte-Taylor, J. 2008. *My Stroke of Insight.* Viking Press.

Booker, C. 2004. *The Seven Basic Plots: Why we tell stories.* Continuum.

Brooks, D. 2012 *The Social Animal* . Random House.

Boyne, A.F., Bohan, T.P., and Williams, T.H. 1974. *The effects of calcium-containing fixation solutions on cholinergic synaptic vesicles.* J. Cell Biol. 63:780-795.

Boyne, A.F., Bohan, T.P. and Williams, T.H. 1975. *Changes in cholinergic synaptic vesicle populations and the ultrastructure of the nerve terminal membrane of Narcine brasiliensis electric organ stimulated to fatigue in vivo.* J. Cell Biol. 67:814-825.

Boyne, A.F. and McLeod, S. 1979. *Ultrastructural plasticity in stimulated nerve terminals: Pseudopodial invasions of abutted terminals in Torpedine ray electric organ.* Neuroscience 4:615-624.

Boyne, A.F. 1979. *A Gentle, Bounce-free Assembly for Quick-freezing Tissues for Electron Microscopy.* J. Neurosci. Methods, 1:353-364.

Boyne, A.F. and Tarrant, S.B. 1982 *Pseudopodial interdigitations between abutted nerve terminals: diffusion traps which occur in several nuclei of the rat limbic system.* J. Neuroscience 2: 463-469.

Broca, P. 1879. Bull Acad Med Paris; 2S:1331-1347.

Calvin, W.H. 1983. *The Throwing Madonna.* McGraw Hill.

Calvin, W.H. 1996 *How Brains Think.* Basic Books.

Calvin, W.H. 1996. *The Cerebral Code: Thinking a Thought in the mosaics of the Mind.*

Chabris, C. and Simons, D. 2011. *The Invisible Gorilla.* Crown.

Damasio, A. 1999. *The Feeling of What Happens: Body and Emotion in the Making of Consciousness.* Harcourt Brace.

De Robertis, E.P.D., & H.S. Bennett, 1955. *Synaptic Vesicles.* J. Biophys. Biochem., Cytol. 1: 47-58.

Dennett, D.C. 1991. *Consciousness Explained.* Boston. Little, Brown.

Domke, D.S. and Coe, K.M. 2008. *The God Strategy: How Religion became a Political Weapon In America.* Oxford University Press.

Dowdall, M.J., Boyne, A.F. and Whittaker, V.P. 1974. *Adenosine Triphosphate: A constituent of cholinergic synaptic vesicles.* Biochem J. 140:1.

Dundy, Elaine. 2004. *Elvis and Gladys.* University Press of Mississippi, Repr.

Flaherty, A.W. 2004. *The Midnight Disease: The Drive to Write, Writer's Block, and the Creative Brain.* New York. Houghton Mifflin.

Frankl, V.E. 1959. *Man's Search for Meaning.* Beacon Press.

Freud, S. 1913. *The Interpretation of Dreams.* The Macmillan Company.

Garey, L. 2006. *Brodmann's 'Localisation in the Cerebral Cortex,'* Springer.

Gazzaniga, M.S. 2008. *Human: The Science Behind what Makes us Unique.* Harper Collins.

Gazzaniga, Michael S. 1985. *The Social Brain: Discovering the Networks of the Mind.* Basic Books.

Geschwind, N. and Galaburda, A.M. 1987. *Cerebral Lateralization: Biological Mechanisms, Associations, and Pathology.* The MIT Press.

Gordon, H.W. and Bogen, J.E. 1974 *Hemispheric lateralization of singing after sodium amylobarbitone.* J. Neurol. Neurosurg.Psychiatry 37:727-738.

Guth, P. S.; Norris, C. H. 1993. *The pharmacology of auditory and vestibular systems.* J. Oto-rhino-laryngology and its related specialties 55(3):180-1.

Hermann L. F., Helmholtz, M. D. 1912. *On the Sensations of Tone as a Physiological Basis for the Theory of Music.* (Fourth edition.) Longmans, Green, and Co.

Heuser, J.E. and Reese, T. S. 1973. *Evidence for recycling of synaptic vesicle membrane during transmitter release at the frog neuromuscular junction.* J. Cell Biol. 57:315.

Heuser, J.E., Reese, T.S., Dennis, M.J., Jan, Y., Jan, L. and Evans, L. 1979. *Synaptic vesicle exocytosis captured by quick freezing and correlated with quantal transmitter release.* J. Cell Biol. 81: 275-300.

Hirsch, E. 2002. *The Demon and the Angel: Searching for the Source of Artistic Inspiration.* Harcourt.

Hofstadter, D.R. 2007 *I am a Strange Loop*. Basic Books

Hume, D. 1748. *Enquiries concerning Human Understanding and Concerning the Principles of Morals*. 3rd ed. L.A. Selby-Bigg (ed.)1975. Clarendon Press, section vii, Part I, p.62.

Ivry, R.B. and Robertson, L.C. 1998. *The Two Sides of Perception*. MIT Press.

Jennings, H.S. 1911. *Behavior of the Lower Organisms*. Repr. Indiana University Press, 1962.

Jaynes, Julian. 1976. *The Origins of Consciousness in the Breakdown of the Bicameral Mind*. Houghton Mifflin.

Kandel, E.R. 2006. *In Search of memory: The Emergence of a New Science of Mind*. New York. Norton Press.

Karnovsky, M.J. 1965. *A formaldehyde glutaraldehyde fixative of high osmolarity for use in electron microscopy*. J. Cell Biol. 27:137A-138A.

Katz, B and Miledi, R. 1965. *The effect of calcium on acetylcholine release from motor nerve terminals*. Proc. R. Soc. Lond. B. Biol. Sci. 161:496.

LaBerge, S. 1990. *Lucid Dreaming: The power of being aware and awake in your dreams*. Ballantine Press.

Lakoff, G. 2008. *The Political Mind: Why you can't understand 21st-century American politics with an 18th-century brain*. Viking Penguin.

Libet, B. 2005. *Mind Time: The Temporal Factor in Consciousness* Perspectives in Cognitive Neuroscience. Harvard University Press.

Linden, D.J. 2007. *The Accidental Mind: How Brain Evolution has given us Love, Memory, Dreams and God*. Belknap Press of Harvard University.

MacNeilage, P.F., Rogers, L.J. and Vallortiga, G. 2009. *Evolutionary origins of your right and left brains*. Scientific American. July.

McGilchrist, Iain. 2010. *The Master and his Emissary; The Divided Brain and the Making of the Western World*. Yale University Press.

Mendeleyev, D. I.; Jensen, W. B. 2005. *Mendeleev on the Periodic Law: Selected Writings, 1869 - 1905*. Dover.

Miller, S.M., Liu, G.B., Ngo, T.T., Hooper G., Riek S., Carson, R.G.,

Pettigrew, J. D. 2000 *Inter-hemispheric switching mediates perceptual rivalry.* Current Biology 10 383-392.

Mithen, S. 2005. *The Singing Neanderthals: The Origins of Music, Language, Mind and Body.* Weidenfeld and Nicholson.

Nisbett, R.E. 2009. *Intelligence and How to Get It.* Norton Press.

Nørretranders, Tor. 1998. The User Illusion: Cutting Consciousness Down to Size. Viking.

Ornstein, R.E. 1972. *The Psychology of Consciousness.* Penguin Books.

Ornstein, R.E.1997. *The Right Mind: Making Sense of the Hemispheres.* Harcourt Brace.

Pert, C.B. 1999 *Molecules of Emotion: The Science behind Mind-Body Medicine.* Touchstone (Simon and Schuster).

Phillips, T.E. and Boyne, A.F. 1984. *Liquid nitrogen based quick freezing: Experiences with bounce free delivery of cholinergic nerve terminals to a metal surface.* J. Elect. Microsc.Technique 1:9-29.

Pinker, S. 2007. *The Stuff of Thought.* Viking Adult.

Politoff, A.L., Rose, S. and Pappas, G.D. 1974. *The calcium binding sites of synaptic vesicles of the frog sartorius neuromuscular junction.* J. Cell Biol. 61: 818-823.

Pribram, Karl. 1971. *Languages of the brain; experimental paradoxes and principles in neuropsychology.* Prentice-Hall.

Ramachandran, V.S. and Blakeslee, S. 1998. *Phantoms in the Brain: Probing the Mysteries of the Human Mind.* William Morrow and Co.

Sachs, O. 2008. *Musicophilia: Tales of Music and the Brain: Revised and Expanded.* Vintage Press.

Sharma, S. Coombs, S., Patton, P and Burt de Perera, T. 2009. *The function of wall-following behaviours in the Mexican blind cavefish and a sighted relative, the Mexican tetra (Astyanax).* Journal of Comparative Physiology. 195, Issue: 3; 225-240.

Shelley, M. 1818. *Frankenstein: The Modern Prometheus.*

Shlain, L. 1998. *TheAlphabet versus the Goddess: The conflict between word and image.* Viking.

Silinsky, E.M. and Hubbard, J.I. 1973. *Release of ATP from rat motor nerve terminals.* Nature, 5407:404-405.

Smith, A. 1776. *An Inquiry into the Nature and Causes of the Wealth of Nations.* University Of Chicago Press, (1977 reprint).

Taylor, J.B. 2008. *My Stroke of Insight.* Viking Press.

Van Harreveld, A and Crowell, J. 1964. *Electron Microscopy after rapid freezing on a metal; surface and substitution fixation.* Anat. Rec. 149:381-386.

Wittkowski, George. 1943 *Swift's Modest proposal: The Biography of an early Gerogian Pamphlet Journal of the History of Ideas,* 4 (1) :75-104.

Wigan, A.L. 1985 (orig 1844) *The duality of the Mind.* J. Simon

Wolfram, S. 2002 *A New Kind of Science.* Wolfram Media.

Zaidel, D. 1990. *Long Term Semantic Memory in the Two Cerebral Hemispheres.* In *Brain Circuits and Functions of the Mind. Essays in Honor of Roger W. Sperry.* Colwyn Trevarthen ed. Cambridge U.P.

Index